T0305509

Emerging Technologies for Sustainable and Smart Energy

Considering the alarming issue of global climate change and its drastic consequences, there is an urgent need to further develop smart and innovative solutions for the energy sector. The goal of sustainable and smart energy for present and future generations can be achieved by integrating emerging technologies into the existing energy infrastructure. This book focuses on the role and significance of emerging technologies in the energy sector and covers the various technological interventions for both conventional and unconventional energy resources and provides meaningful insights into smart and sustainable energy solutions. The book also discusses future directions for smart and sustainable developments in the energy sector.

Prospects in Smart Technologies
Series Editors

Mohammad M. Banat
Jordan University of Science and Technology
Irbid, Jordan

Sara Paiva
Instituto Politécnico de Viana do Castelo
Viana do Castelo, Portugal

Emerging Technologies for Sustainable and Smart Energy
Edited by Anirbid Sircar, Gautami Tripathi, Namrata Bist, Kashish Ara Shakil and Mithileysh Sathiyanarayanan

For more information on this series, please visit: www.routledge.com/Prospects-in-Smart-Technologies/book-series/CRCPST

Emerging Technologies for Sustainable and Smart Energy

Edited by
Anirbid Sircar
Gautami Tripathi
Namrata Bist
Kashish Ara Shakil
Mithileysh Sathiyanarayanan

CRC Press is an imprint of the
Taylor & Francis Group, an **informa** business

First edition published 2023
by CRC Press
6000 Broken Sound Parkway NW, Suite 300, Boca Raton, FL 33487-2742

and by CRC Press
4 Park Square, Milton Park, Abingdon, Oxon, OX14 4RN

CRC Press is an imprint of Taylor & Francis Group, LLC

ISBN: 978-1-032-30428-1 (hbk)
ISBN: 978-1-032-30949-1 (pbk)
ISBN: 978-1-003-30740-2 (ebk)

DOI: 10.1201/b23013

Typeset in Times
by SPi Technologies India Pvt Ltd (Straive)

Contents

Preface

The journey of this book from inception to its present form has been, like every journey, an interesting one. A beautiful amalgamation of technical work by academicians and industry personnel on the topic of emerging technologies for sustainable and smart energy is presented in this book. With the advent of COVID-19, the world has relied heavily on the inclusion of technology with our day-to-day lives. Considering the alarming issue of climate change and its drastic consequences, there is an urgent need to develop technology-based smart and innovative solutions for sustainable development in the energy sector. Today's world is heavily focused on nonrenewable energy for its energy requirements. Thus this book focuses on the applicability of technological interventions in the area of smart and sustainable energy to address the world's increasing energy demands. This book also throws light on inclusion of both renewable and unconventional energy sources as the backbone of the energy needs of the world. Interesting examples of renewable and sustainable nonrenewable energies in the form of the book chapters have been presented. The book also discusses connecting the dots between unconventional energy and renewable energy with present- and future-day technologies for a sustainable future. Further the book covers the sustainability aspects of the integration of technology into the energy sector and the development of sustainable solutions for future energy solutions and smart cities. This book further compliments the current global need for sustainable energy solutions and also helps to understand future energy requirements.

Since smart energy encompasses several domains, the book provides a multidimensional view for the masses ranging from computer science, civil engineering, mechanical engineering along with environmental and social science. The book covers areas such as inclusion of artificial intelligence and the Internet of Things in everyday energy requirements, technology for forecasting the energy needs for the future, etc. The book deals with the concepts of smart cities, emerging technologies and sustainable energy among others. The various opportunities, challenges and issues are also covered. The editors are hopeful the book shall provide a foundation for researchers, scientists and academicians working in the area of sustainability and the smart cities ecosystem in the energy sector.

Anirbid Sircar
Gautami Tripathi
Namrata Bist
Kashish Ara Shakil
Mithileysh Sathiyanarayanan

Editors

Anirbid Sircar, PhD, is a petroleum professional with 27 years of experience in the field of academics and industry. Professor Sircar earned an MTech in petroleum exploration and a PhD in reservoir tomography at Indian Institute of Technology (Indian School of Mines), Dhanbad. He has developed expertise in exploration, natural gas processing and transportation. Professor Sircar has supervised ten PhD students and 36 master's students. He has completed several consultancies and MDPs with industries. He has authored three books on city gas distribution and two books in geothermal energy. Professor Sircar has won several awards at the national level, such as Indira Gandhi Shiromani award, Bharat Vikash award and Venus award for distinguished faculty. He is highly experienced in the energy sector, and his expertise has helped in identifying areas in the energy sector where technological interventions can be helpful. Also, his rich experience helps in exploring the various aspects of renewable and nonrenewable energy resources and inviting the best researches in the area.

Gautami Tripathi is an Assistant Professor in the Department of Computer Science and Engineering, Jamia Hamdard, New Delhi, India. She has five plus years of teaching and research experience. Apart from her teaching duties, she is also involved in various research activities and important departmental committees. Her research interests lie in the area of the Internet of Things, blockchain, big data, data science and sustainable development. Currently she is working in the area of IoT and Blockchain technology. She has supervised over 40 undergraduate and postgraduate students. She has several publications in reputed national and international journals and conferences and is on the editorial board of many reputed journals. She is also the reviewer for many reputed journals and the member of the organizing committee for several conferences and workshops. She is also a member of reputed professional bodies such as IEEE and ACM and has delivered many special lectures and presentations in several workshops across India. She is the Co-PI of a government-funded project under the scheme to fund for improvement of S&T infrastructure (FIST) of the Department of Science and Technology (DST), Government of India. She is associated with many NGOs and actively work in the area of rural development using technology. Gautami Tripathi has experience in emerging technologies and has also worked in the area of smart cities. Her expertise helps to identify the best feasible technology-based solutions for smart and sustainable energy.

Namrata Bist is an Assistant Professor in the School of Petroleum Technology, Pandit Deendayal Petroleum University, Gandhinagar, Gujarat, India. She has rich work experience from reputed firms such as Weatherford International and Infosys Ltd. She is pursuing a PhD in hybrid renewable energy. She has a keen interest in various research areas such as city gas distribution, renewable energy and hydraulic fracturing. Her teaching specializations are production engineering, unconventional hydrocarbon resources, natural gas engineering and safety, health and environment.

She is actively involved in organizing and conducting soft skills development programs for students and industry professionals. She is a passionate and efficient professional who handles teaching, research and administration work in a balanced manner. She has good experience in hybrid renewable energy and city gas distribution. Her expertise helps to identify and explore hybrid energy solutions for smart city infrastructures and in inviting the best research in the area.

Kashish Ara Shakil, PhD, earned a PhD in computer science at Jamia Millia Islamia, New Delhi. She is an Assistant Professor in the College of Computer and Information Sciences, Princess Nourah Bint Abdul Rahman University, Riyadh, Saudi Arabia. She earned a bachelor's in computer science at Delhi University, New Delhi, India, and an MCA degree at Jamia Hamdard University, New Delhi, India. She serves as the co-editor in chief of *Journal of Applied Information Science* and is on the editorial boards of many reputed international journals in computer sciences and has published several research papers. She has written two books, *Internet of Things (IoT): Concepts and Applications (S.M.A.R.T. Environments)* and *Green Automation for Sustainable Environment*. Her areas of interest include database management using cloud computing, big data analytics, IoT, distributed and service computing. Dr. Kashish has a rich experience in the IoT domain, which is one of the key concepts in developing smart infrastructures. Her experience in the area of sustainable environments helps to identify sustainable solutions for the energy sector.

Mithileysh Sathiyanarayanan, PhD, is an Indian-born research scientist and entrepreneur based out of London. He works closely with "Make in India" initiatives. His expertise in technology research at both the theoretical and application levels ranges from industrial projects to academic ventures in domains such as education, engineering and healthcare. He earned a PhD at the City University of London and a predoctoral fellowship at the University of Brighton, UK. His research was in collaboration with Nokia Research, Finland, which won him a Young Scientist award. Dr. Sathiyanarayanan has won several research and entrepreneurship awards and has been invited as a research consultant at various industries for his research excellence.

Contributors

Jose Anand
Department of Electrical and Computer Engineering
KCG College of Technology
Chennai, India

Anusha
School of Computer Science and Engineering
Vellore Institute of Technology (VIT)
Vellore, Tamil Nadu, India

Namrata Bist
School of Petroleum Technology
Pandit Deendayal Petroleum University
Gandhinagar, Gujarat, India

Sayandeep Chandra
Tata Consultancy Services, India

Geetanjali Chauhan
Department of Petroleum Engineering
Indian Institute of Petroleum & Energy
Visakhapatnam, Andhra Pradesh, India

Swapnil Dharaskar
Department of Chemical Engineering
School of Technology
Pandit Deendayal Energy University
Gandhinagar, Gujarat, India

Kaushalkumar Dudhat
School of Petroleum Technology
Pandit Deendayal Energy University
Gandhinagar, Gujarat, India

Vishweash Gurjar
Department of Electrical Engineering
School of Technology
Pandit Deendayal Energy University
Gandhinagar, Gujarat, India

N. V. Haritha
Department of Electrical and Electronics Engineering
Meenakshi Sundararajan Engineering College
Chennai, India

Jayashree
School of Computer Science and Engineering
Vellore Institute of Technology (VIT)
Vellore, Tamil Nadu, India

Siddharth Joshi
Department of Electrical Engineering
School of Technology
Pandit Deendayal Energy University
Gandhinagar, Gujarat, India

Meera Karamta
Department of Electrical Engineering
School of Technology
Pandit Deendayal Energy University
Gandhinagar, Gujarat, India

Sunil Kumar Khare
Department of Petroleum Engineering and Earth Sciences
University of Petroleum and Energy Studies
Dehradun, Uttarakhand, India

Sujay Kore
Department of Chemical Engineering
School of Technology
Pandit Deendayal Energy University
Gandhinagar, Gujarat, India

Subhankar Mazumdar
Tata Consultancy Services, India

Payal Mehta
Department of Information Technology
Institute of Technology
Nirma University
Ahmedabad, Gujarat, India

Saurabh Mishra
Centre for Advanced Studies
Lucknow, India

Ayushi Rawat
Trane Technologies
Bengaluru, Karnataka, India

Kamakshi Rayavarapu
Pandit Deendayal Energy University
Gandhinagar, Gujarat, India

Jaimin Shah
Department of Electrical Engineering
Faculty of Technology and Engineering
The Maharaja Sayajirao University of Baroda
Vadodara, Gujarat, India

Manan Shah
Department of Chemical Engineering
School of Technology
Pandit Deendayal Energy University
Gandhinagar, Gujarat, India

Vidhi Shah
Department of Electrical Engineering
School of Technology
Pandit Deendayal Energy University
Gandhinagar, Gujarat, India

Vrutang Shah
School of Petroleum Technology
Pandit Deendayal Energy University
Gandhinagar, Gujarat, India

Anirbid Sircar
School of Petroleum Technology
Pandit Deendayal Petroleum University
Gandhinagar, Gujarat, India

Sugat Srivastava
Presidency University
Bengaluru, India

Yash Thakare
Department of Chemical Engineering
School of Technology
Pandit Deendayal Energy University
Gandhinagar, Gujarat, India

Vijayashree
School of Computer Science and Engineering
Vellore Institute of Technology (VIT)
Vellore, Tamil Nadu, India

Kriti Yadav
UNESCO, GRO-GTP Fellow and Research Associate
Centre of Excellence for Geothermal Energy
Pandit Deendayal Petroleum University
Gandhinagar, Gujarat, India

Mohamed Yousuff
School of Computer Science and Engineering
Vellore Institute of Technology (VIT)
Vellore, Tamil Nadu, India

1 Emerging Technologies in Conventional and Nonconventional Energy Sources

Yash Thakare, Sujay Kore, and Swapnil Dharaskar
Pandit Deendayal Energy University, Gandhinagar, India

CONTENTS

DOI: 10.1201/b23013-1

1.1 INTRODUCTION

Energy is an important component of economic infrastructure. It is deeply linked to the history and development of mankind. The proportion of energy that is being utilized by the community is an indicator of their efficient progress. Significant force in economic, political, social and environmental aspects has become one of the most discussed issues around the world [1]. Energy possesses a very crucial role in our existence; it contributes extensively to our comfort zone, boosts productivity and also permits us to live the manner we wish to. Right from the origination of humankind, we have been exploiting energy sources such as water, wood and fossil fuels as for heating and working machines. We rely on some force for almost all types of activities [2]. Exploitation of various energy sources has been determining the progress of community since ages. In developing countries like India, the exigency for the energy from various fields such as industries, agriculture, housing and the economy is ever increasing. Desire of the energy in the backcountry is constantly increasing. India is the massive consumer of the energy across the globe and stands fourth in terms of consumption. Current energy consumption is almost in the purpose of lighting and cooking, transportation, tillage, farming and commercial areas. The industrialization and technological development of the world has brought high energy requirement to the whole world. India's energy basket is a combination of all available reservoirs which also take in renewable energy as an important resource. Mankind needs a lot of energy resources to survive. The importance of energy in the world context is steadily increasing thereby interaction with the public as well as environment is becoming further apparent. Since the linkage between the growth and progress with the energy utilization and availability is known, the poverty can be combated by putting an end to the shortage of the energy, thereby escalating the access to the advanced energy services connecting development segments to billions of poor people [3–5]. Lack of access to energy strongly influences opportunity.

Reserves of conventional energy sources vary from country to country [6]. Therefore, it led to major environmental concerns, serious political conflict, inevitable economic dependence and significant social consequences. The current situation and future predictions for energy needs motivate people to search for alternative energy sources. In addition, current and future potential environmental, economic, political and social adverse developments will also force countries to lean towards renewable energy sources. In this regard, emerging energy technologies have become the answer to sustainable energy planning [7]. Energy decides the development of the human society. Currently, the public across the globe should look forward towards the advantages of the advanced energy services that are being provided to everyone and are handed over as purely, securely and systematically. The major constrain for worldwide economy is to switch to the energy sources that are sustainable in nature and at the particular instance think of replacing our existing models of the energy that are purely dependent on the decreasing resources [8, 9]. Availability, reliability, equity and cost are the basic elements to develop or execute any plan-relating steady energy. Novel approaches need proper technology, proper commercialized models in order to impart power and good time to the customers, and an appropriate and compatible scheme structure along with the essential dedication of local organizations. This chapter gives you an idea of the many emerging technologies and their importance in conventional and nonconventional energy sources.

1.2 ENERGY SOURCES CLASSIFICATION

Sources of energy are classified into two main categories which are grounded on how swiftly the sources can freshen:

- Conventional energy sources
- Nonconventional energy sources

1.2.1 Conventional Energy Sources

Conventional energy sources can also be termed as renewable energy sources. They are limited in terms of quantity and have been used by humans for a very long time. These sources of renewable energy are deteriorating materials that take centuries to compose [10]. Few examples of it can be petroleum products, coal and many more. Therefore, once these exhausts, they can never be produced at speed or at any cost, so as to sustain the consumption rate. Most of these resources are depleted due to unceasing utilization [11]. It is taken under consideration that the petroleum reserves of our country will run out in some years and reserves of coal will continue for another few more decades. Few examples of conventional sources of energy are petroleum, coal, electricity and natural gas.

The conventional energy sources then can be even more divided into two important categories: one is the commercial sources of energy and the other is the noncommercial sources of energy.

1.2.1.1 Commercial Sources of Energy

The commercial sources of energy are such energy sources where the user has to pay in order to make use of it. Examples for such sources are petroleum coal, electricity, natural gas and oil.

1.2.1.1.1 Coal

Without any secondary thought, coal is actually the most important energy source that will remain for more than two centuries. It is basically a sedimentary rock which has brown—black appearance. It gets developed when the perished plant material gets converted to peat (incompletely degraded organic matter or vegetation accumulation) that is transformed to the coal with the temperature and pressure for many millions of centuries. Percentage of carbon present in the coal is very high. It also contains some traces of other elements such as oxygen, nitrogen, hydrogen and sulphur. Coal is formed when plant residues are converted into lignite and later into anthracite. It is a very long operation and requires a very extensive period of time. Coal can be utilized in various proposals like electricity generation via thermal power plant, steam engines as fuel, household purpose and many more [5, 12]. Around 70% of the country's total consumption of energy is accounted by the coal itself. India has more than 148787 coal reserves. India is the fourth largest administrator of the coal.

1.2.1.1.2 Oil and Natural Gas

Oil is regarded as black gold in the liquid state, whereas natural gas is also an essential energy source across the globe. The oil is made from a tremendously huge number of small plants and animals like when they perish, they are caught under several layers which are of sand and soil on the seabed and subjected to high temperature and stress [13]. It has wide applications in the various modes of transport as a fuel. Natural gas whereas is composed when many blankets of decaying animals and plants under the earth's surface are subjected to extreme temperature and pressure for a very long time. It can be used for a variety of purposes including cooking, heating and power generation. It causes less air pollution than other fossil fuels [14].

1.2.1.1.3 Electricity

It is the most ordinary energy form and is utilized for both household and business motive [15]. It can be generated by utilizing the sources of energy such as fossil fuels, renewables and nuclear power. Electricity is mainly used in various electrical gadgets which incorporate television, fridge, microwave, grinder and many more.

The two major sources for generating power are mentioned below:

- Nuclear power
- Thermal power

1.2.1.1.3.1 Nuclear Power Elements such as plutonium and uranium are widely utilized as a source of fuel in the nuclear power plants. These are inexpensive when compared to that of coal. Most of the electricity generated is due to the three key cases which are nuclear fusion, nuclear fission and the reactions due to nuclear decay. A very minute quantity of radioactive material produces a huge amount of energy [16]. In order to acquire nuclear power, it is must that nuclear reactions should take place. Coal produces less greenhouse gas emissions during power generation than sources such as power plants because nuclear power is one of the most eco-friendly conventional sources of energy [17].

1.2.1.1.3.2 Thermal Power Thermal power is generated in many power plants using coal and oil. Thermal power generation means converting fuel into heat. This is engendered employing thermal generators in the thermal power plant [18]. This plant generally burns the fuel in order to boil the water and produce steam. This generated steam is then circulates to turbine which is connected to an electric generator. The transmission of electricity is way more systematic than conveying petroleum or the coal for the exact interspace [19].

1.2.1.2 Noncommercial Sources of Energy

Nonconventional sources of energy are basically that kind of sources which are present in the nature free of charge and can be exploited at ease. Firewood, cow dung, dry leaves and grass are few examples of it.

1.2.1.2.1 Fuel Woods

The public in village use fuel wood for their daily food preparation. It comes from the farms and forest. The accessibility of fuel wood these days has become tough owing to fast desertification [20, 21]. This issue can be resolved by the afforestation.

1.2.1.3 Advantages of Conventional Sources of Energy

Upcoming are few advantages of conventional energy sources:

- More efficient
- Less production cost

1.2.1.4 Disadvantages of Conventional Sources of Energy

Upcoming are few disadvantages of conventional energy sources:

- Non-eco-friendly
- Can be depleted

1.2.2 Nonconventional Sources of Energy

Nontraditional sources of energy are those kinds of sources which are continuously freshened by natural phenomenon. These cannot be easily removed, can be produced continuously and can be used repeatedly. Energy derived from nonconventional sources such as wind energy, solar energy, wave energy, geothermal energy and biomass energy are called nonconventional energy [22, 23]. This resource doesn't harm the environment and also have very less expenses. They are called renewable resources because they are replaced by natural operations at a rate which is equal to or even greater than their consumption rate [24].

The various types of nonconvention sources are as follows:

- Solar energy
- Wind energy
- Tidal energy
- Geothermal energy
- Biomass

1.2.2.1 Solar Energy

Sun is a primary source of cleansed and continuous energy. Sun cannot be expired, but it can be maintained almost every day as the sun shines; hence, it is a sustainable source of energy [25]. In solar plants, it is used by direct conversion of solar energy into electrical energy. Depending upon energy form to be generated, photovoltaics are exposed to the sun. This energy is an important energy source but its consumption is too low. It is an essential nonconventional source of energy and provides pollution free, environmentally friendly product and is available in abundance [26].

1.2.2.2 Wind Energy

Wind energy is generated using wind power. This is extensively employed in the functioning of the water pumps with the motive of irrigation. Wind is natural movement of the air. Air is utilized to rotate windmill blades that twist the shaft and this motion is provided by the generator or the pump generates electricity [27]. When the momentum of the wind boosts, the output for the power also boosts to the highest output of a specific turbine. Burning fossil fuels to generate electricity eliminates pollution.

1.2.2.3 Tidal Energy

Tidal power is generated with the use of waves of the ocean. As a renewable energy source, tidal power has not yet been used owing to its expensive technology. Tides occur every 12 hours because of the gravitational pull of the moon. Alike hydroelectricity produced from dams, tidal water is trapped in a weir during tidal flow and enforced across hydro-turbines during low tides. Investment in tidal power plants is huge [28]. So in order to get enough energy from the tidal force, the peak of the high tides has to be at least sixteen feet elevated than the low tide.

1.2.2.4 Geothermal Energy

Geothermal power is the thermal energy that is obtained from the hot rocks present under the surface of the Earth. Geothermal wells emit trapped GHGs into the earth's surface [28]. This release is very less per unit of energy when compared to that of fossil fuels. The energy obtained usually has lower cost because it cut back 80% of fossil fuels. This is the reason which has led to an incremental utilization of geothermal energy. It assists in reducing pollution and restricts negative changes in climate.

1.2.2.5 Bio Energy

Biomass is a natural depend crafted from wood, plants, sewage and animals [29]. These substances burn to generate heat energy that is then used to generate electricity. Plants restore sun power via photosynthesis to supply biomass. This biomass generates distinctive forms of power reasserts via distinctive forms of cycles. The chemical composition of biomass varies from species to species; however, biomass usually incorporates 75% carbohydrate, 25% lignin, or sugar. Biomass power is likewise implemented to lighting, cooking, and power generation and the leftover after biogas extraction are a good source of manure [30].

1.2.2.6 Advantages of Nonconventional Energy Sources

Upcoming are few advantages of nonconventional energy sources:

- Environmentally friendly
- Inexhaustible
- Easy to operate

1.2.2.7 Disadvantages of Nonconventional Energy Sources

Upcoming are few nonconventional energy sources:

- Low-efficiency levels
- High cost
- Inconsistent, unreliable supply
- Harmful to wildlife and surrounding environment

1.3 EMERGING TECHNOLOGY IN CONVENTIONAL ENERGY SOURCES

This section presents the emerging technologies in the conventional energy sources.

1.3.1 Advanced Nuclear Reactors

These days about 440 nuclear power reactors with a complete hooked up capability of 350 GWe offer 16% of the world's power. Power generated from the nuclear plants is in general usage in various parts across the globe [31]. Till now the mentality was just to build nuclear power plant with gradually growing capacity degrees in order to make profit from the monetary advantages of the large scale. In the past few decades, however, investment in smaller, modular units has multiplied. The benefit of those plant life is that much of the development work takes area in factories, with best a small component carried out on the site, and this could significantly decrease each production time and capital cost [32]. With the modern and simple reactor designs, these days' nuclear energy stations operated as base-load gadgets can be competitive to traditional power plants.

For the reactors operating at high temperature, improvement focuses in particular on so-known as pebble-bed reactors. Design of this is such that the fuel is encircled in the balls made up of graphite, to which around 15000 minute balls of uranium dioxide are present along with the coating of thick graphite. The reactor may also include a lot of those balls of the fuel, in the internal of a metal pressure vessel that is encircled by means of a graphite block reflector. Also approximately thousands of unfuelled balls made of graphite are packed into the centre to manipulate its temperature distribution by using spacing out the recent gas balls. Extreme temperatures can also be withstood by the balls as they contain refractory materials.

The key benefits of these reactors are as follows:

- More thermal efficiency
- High reactor temperature
- The materials are capable of withstanding extremely high temperature.
- Less hazardous and also prevents the accidents arising from the coolant loss.
- Interaction of the other materials with the coolant present in the reactor is prevented.

Most nuclear reactor designs are capable of utilizing just 1% of the raw nuclear energy source to produce fuel but the fast breeder reactor has the sole capability of consuming the entire nuclear energy source to generate power. The primary advancements in fast breeder reactors are as follows:

- The utilization of the molten elements like lead and bismuth is being done as a refrigerant rather than that of the molten sodium metal. These chemicals have zero reactivity with the water unlike sodium metal; hence, there will be no any reaction taking place thus thereby minimizing the hazard [33].
- Advancement of the swift fission systems to alter extensively resided trans-uranic nuclides into the by-product which have small half lifetime.

1.3.2 Fusion Energy

Nuclear fusion—the formation of large nuclei by the accumulation of light atomic nuclei—is likely to generate huge amounts of energy with remarkable eco-friendly benefits [34]. The primary benefit of energy obtained as a result of the fusion is that it is a rationally never-ending source of energy and does not produce any carbon dioxide.

Energy obtained as a result of fusion is dependent on a combination of two important isotopes of the hydrogen element. They are namely deuterium and tritium. Currently, enormous stockpile of these isotopes is available; the required quantity at the Fusion Power Station is very small.

Fusions create neutrons that are absorbed through the fuel chamber walls; radiating materials of the wall and alter its characteristics. The outcome of the present material progress studies shows that the radioactivity of the waste generated from the fusion decreasing at a faster rate; after a century later this generated waste will be having even low radioactivity than that of the generating from thermal power plants [35]. Yet, the waste obtained as a part of fusion will be an issue to be resolved.

At the elevated temperatures which are required for the fusion, the fuel becomes plasma which is entirely an ionized gas. Powerful magnetic fields are being used by the magnetic constraints so as to maintain the hot plasma apart from the walls of the compartment wherein the reaction takes place. This compartment has a shape of doughnut and produces a strong field of magnetism as an enclosed hoop that squeezes the plasma inside the ring, as plasma does not meet the compartment walls, the magnetic incarceration impart efficient thermal insulation at extremely elevated temperatures of around or more than 100 million Celsius per minute. Required fields of the magnetism are created by the electricity that flows in the inside of the coils that are present outside of the plasma compartment. Many researches on the fusion generally utilize inner currents inside the plasma and the outer currents in the magnetic coil. One disadvantage of this advancement is that it is tough to keep inner current without any disturbance, since they are very effectively produced by alternation of the external current by using the transformer. There is a different positioning known as stellar device which is convenient to use continuously owing to its sole reliability on the persistent external currents present in the magnetic coil.

The fusion technique, called inertial confinement, gives a substitute to the magnetic confinement of the plasma. In inertia capture, fuel prills with a diameter of a few millimetres are brightened with the high-power laser or particle beam. Up to about a 10^{-9} of a second, every fuel cartridge is heated sufficiently to cause compression and coupling reactions. After these billion seconds, the inertia of the pellets does not have the energy which was provided by the beam of radiation thus resulting in the successive fusion reaction ultimately leading to the bullet explosion. A power plant which utilizes the inertia restraint processes fuel prills at a constant charge of 10–20 per second. In many nations across the globe, the military budget supports research on inertia detention and is partly classified as related to nuclear weapons.

1.4 EMERGING TECHNOLOGY IN NONCONVENTIONAL ENERGY SOURCES

1.4.1 Solar Energy

Solar energy is potentially efficient in supplying of the chilling, electricity and heating needs to the world. Development in technologies in solar energy sources have been contributing to use of natural energy sources and converting into energies for betterment by solving environmental problems and providing employment as well [36]. The growing technologies provide the use of solar energy converting directly converting sunlight into electricity; photo-catalytic oxidation for clean-up of drinking water, industrial wastewater, air soil and many more heating and cooling technologies some of which are been discussed below.

1.4.1.1 Photovoltaic Cells

Direct conversion of solar energy to the electricity is possibly performed by utilizing photovoltaic cells. This effect of cells depends on photonic interaction with the energy that is more than or equal to the photovoltaic material's band-gap [37].

The first solar cell was constructed by Charles Fritts in the 1880s. Many developing countries are on the verge to adopt solar energy advancements for domestic electricity production using PV cells. Various photovoltaic cell technologies thus can be witnessed in Figure 1.1.

Many researches and advances in technologies are emerging in PV cells. Recently the organic—inorganic halide perovskite is proven efficient for boosting the proficiency of the solar cell in a faster manner. The rise in efficiency of PV cells from 3.8% to 22.1% is seen between years of 2009 and 2016. More efficacy, pliability, and cell structure of the emerging hybrid halide perovskite have attracted most of the scientists and researchers [38].

1.4.1.2 Concentrated Solar Power Technology

The Concentrated Solar Power technology (CSP) provides generation of electricity, by using heat that is produced by the solar irradiation, concentrating them into a smaller area. In CSP technology, lenses or mirrors are used to reflect the sunlight to the receiver, whereas the primary circuit i.e. carrier of thermal energy accumulates the heat and it is then later utilized by the secondary circuit or can be directly be sent to turbine to produce the electricity [39]. Between years 1984 and year 1991 at California US, the first Concentrated Solar power plant was constructed where there was no use of the thermal storage [40].

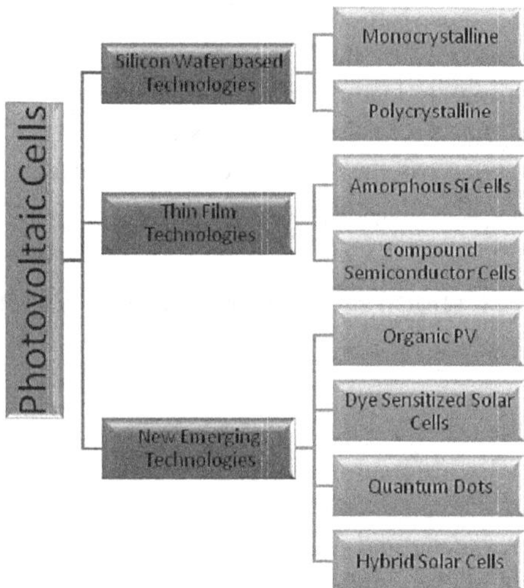

FIGURE 1.1 Advancements in photovoltaic cells.

Since the decade, for the development of the future ages, the CSP technology and the research and development-oriented projects have started and various initiatives have been launched by countries and regions such as US, Australia, Asia and Europe. The projects with lower cost of electricity (LCOE) with higher efficiency technology is been focused. In the year 2011, the US Department of Energy (DOE) Solar Energy Technologies Office (SETO) has launched the SunShot Initiative so as to make solar-generated electricity competitive across the country by year 2020 and provided the funds to theme relating the SunShot's initiative of Gen3 CSP program in 2018. In year 2012, the ASTRI was started by ARENA to improve and evolve these technologies to the next generation [41].

1.4.2 Wind Energy

The evolution from windmills to wind turbines did not happen over the night. Since the beginning of the century, attempts were done to produce the electricity from windmills. Denmark was the first to erect the batch of steel windmills that was specially built to generate the electricity. Amongst the conventional energy sources, wind energy is rapidly growing as forefront and fierce with conventional sources of energy [42].

Since the last decade, significant growth in the wind energy is seen. In year 2015, the Global Wind Energy Council (GWEC) has reported a massive expansion in wind energy projects across the globe. Globally in 2015, growth up to 432.9 GW in installed wind power capacity was reported which represents an increasing growth in the market higher than 17%. Northern America up to88.749 GW, the Caribbean and Latin America of 12.2 GW, Pacific region of 4.8 GW, the Middle East and Africa 3.489 GW and Asia contributed 175.8 GW and in the growth globally [43] Year 2020 is known the best year in the history for the growth with 93 GW new installed which is 53% yearly increase in Wind Industry reported by Global Wind Energy Council

(GWEC). These newer installations of 93 GW increase the global wind power capacity up to 743 GW. Compared to year 2019, an increment of 59% was seen with installation of 86.9 GW in offshore market [44].

There is a huge advancement in technology seen in wind energy industry in terms of electric generators, integration power systems, aerodynamic design, power electronic converters, control theory and mechanical systems. In the operation of wind energy conversion systems (WECS), electric generators and the power electronic converters are two main components in the perspective of electrical engineering [45]

1.4.2.1 Wind Turbine Generator

In terms of rotational speed, wind turbine generators can be classified as variable speed generators, fixed speed, and the limited variable speed generators. Along with these, other wind generators — BDFIG and BDFRG concepts are been introduced recently.

Because of the lack of brush gear and slip rings, the BDFIG—Brushless Doubly fed induction generator—are more reliable and robust.

An emerging technology, the brushless doubly-fed reluctance generator (BDFRG), is considered as a substitute to the live solutions for wind power applications. It emerges in design when compared to the BDFIG, due to existing of a reluctance rotor, which is an iron rotor without copper windings and which is cheaper than the wound rotor or permanent magnet rotor.

Another type of generator in these systems is the Switched reluctance generator (SRG). It is simple in construction and there are no permanent magnets or conductors in the rotor, which reduces its cost and it is with a control system that permits quick changes in the control strategy hence boosting its performance. This operates as a generator, when the windings of the stator are energized and the salient poles of the rotor are away from their aligned position because of the rotating motion of the prime mover. [46]

1.4.3 Bio-Energy

Bioenergy is one of the diversified energy resources that are available to meet the demand for energy. Bioenergy is a renewable energy form that is useful for transport fuel production, heat and electricity generation. This energy uses the Biomass that is derived from the recent living organic materials.

1.4.3.1 Bio-Fuels

Biofuels are basically fuels that are produced using the biomass resources by using eco-friendly approach and recently gained much attention from researchers and scientists worldwide [47]. Presently, various biofuels are produced from biomass. That may be gaseous and liquid forms of fuels. Ethanol, Biodiesel, methane, methanol, Fischer—Tropsch H2 and bio-oil are some of the examples. Globally, the biofuels from various bio-resources using newer biological processes and developing technologies are rapidly increasing. Much advancement in researches is happening in technology for producing biofuels as the new energy source. On the basis of used biomass sources, Biofuels are usually classified as first generation, second generation, third generation, and fourth generation of biofuels which can be also be witnessed in Figure 1.2.

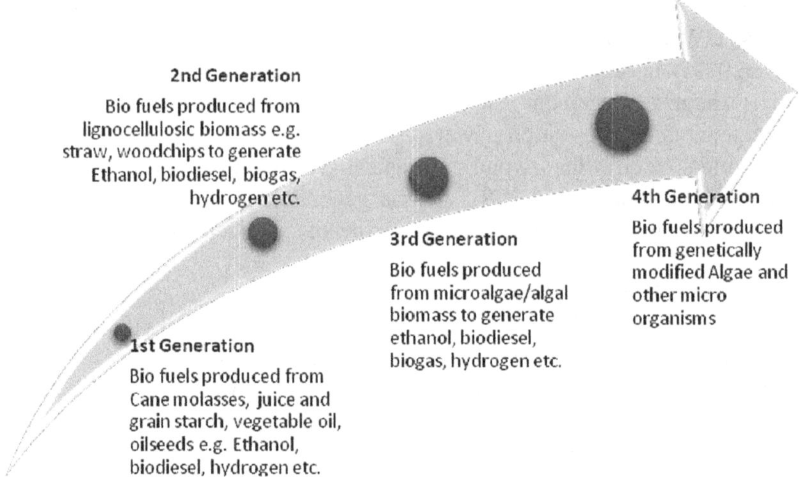

FIGURE 1.2 Recent enhancements in biofuel production.

The 1st generation includes Bioethanol and Biodiesel, which were extracted from the food crops resources such as potato, sugarcane, corn, oilseed, wheat, barley, sunflower soybean [48]. Where use of enzyme—fungal mycelia in fermentation of raw corn and sugarcane was done and Ethanol was produced and was reported as the first biofuel chemical energy [49]. The same results were reported by fermentation of raw cornflour with starch-digesting microbes such as *Rhizopus* sp. and *Saccharomyces cerevisiae*, producing ethanol. From starch, using the initial enzymatic hydrolysis method, large amount of bio-ethanol was produced in the first generation [50]

Second-generation biofuels refer to one which is produced from lignocellulosic materials and various organic waste materials (i.e., wood, straw, switchgrass, etc.) which are easily accessible [51].

Third generation includes the biofuels produced from algae as feedstock and which yields a significant quantity of lipids that produces biodiesel as well as biofuels. Whereas the production of fourth generation of biofuels are depended on the genetically modified organisms as well as modified metabolism routes that provide the greater ability of CO_2 fixation, and the post-genome technology of the microalgae [48].

1.4.4 Ocean Thermal and Tidal Energy

The oceans and seas cover almost 75% of the planet earth, which can potentially make a remarkable contribution to match up the need for energy. The generation of electricity from waves, ocean current and tides is environment-friendly [52]. To make it practically and commercially viable, technologies are been developed to harness wave power, tidal power and ocean thermal energy.

1.4.4.1 Ocean Thermal Energy Conversion (OTEC)

OTEC is a type of renewable form of energy source that is based on the change in temperature of sea, concerned with depth. The temperature gradient in sea is used to operate a thermal machine that produces useful work, which can be further transformed into electricity. We know the ocean captures the heat that is generated by the solar radiation and is covered the earth more than 70%. This ultimately makes Ocean Thermal Energy Conversion systems, a limitless energy source while they are only dependent up on the sun and ocean currents; this productively makes OTEC the huge functional energy storage systems throughout the world. Without harming the environment, power collection ranging from 3 to 5 TW is been estimated [53]. This transformation of the thermal energy into electrical energy is possible using the Rankine cycle i.e. a thermodynamic cycle in which the heat consumption is related to the production of work, in which a liquid evaporates to pass through a turbine. This cycle may be open, closed, or hybrid [54].

1.4.5 Hydrogen and Fuel Cells

Both in Hydrogen cells and Fuel Cells, the electrochemical reaction between oxygen and hydrogen gases produces the electricity. These cells are efficient and reliably applicable for automotive applications. Also for fuel cells, Hydrogen is the chief fuel. Fuel cells can be potentially used widely once they become commercially viable. Hydrogen fuel can be generated from water-electrolysis from the solar energy as well as by extracting from natural gas, sewage gas, biogas or naphtha. Advanced researches are taking place in field of fuel cell including Ammonia fuel cells [55], proton-exchange membrane fuel cells [56] microbial fuel cells with the increasing applicability and adaptability.

1.5 FUTURE SCOPE AND CONCLUSION

Energy is the inescapable requirement of mankind, urging for the newer technologies in the energy sector. With rapid industrialization and a growing population, the demand for energy is constantly increasing. Mankind is making extensive use of these conventional energy sources and therefore the resources are declining more rapidly as the amount of consumption is very high. So in order to fulfil the current demand of energy for the increasing community, researchers are finding out emerging ways to use various developments not only in renewable but also in pollution-free energy sources. The recent techniques in these sectors are partially efficient, but there's lot more techniques that can be effectively looked forward as the emerging techniques for energy production. Looking forward for the safer and efficient energy techniques, researches in the direction of higher efficiency fulfilling energy need of a rapidly growing population is the major challenge forward in the research.

This chapter introduces researchers to the current state of emerging technologies at the forefront of conventional and nonconventional sources of energy. Comprehensive study has been done in this chapter for the several emerging technologies in both conventional and nonconventional energy sources. Advanced technologies for energy generation like advanced nuclear reactors and fusion energy is discussed in detail along with many emerging technologies in the renewable sources of energy like solar, wind, bio, ocean & thermal and hydrogen & fuel cell is also discussed which covers Advanced photovoltaic, Concentrated Solar Power technology, new generation wind turbine generators, emerging biofuel production and Ocean Thermal Energy Conversion. Current research and development status can also be witnessed for these emerging technologies which are evaluated in reference with the global perspective.

With this wide variety of technologies, the globe will overcome energy crisis and also many obstacles for sustainable energy generation. Furthermore, the progression rate for all these technologies is very high that it will fulfil the energy demand for the current population.

REFERENCES

1. A. Brew-Hammond, "Energy access in Africa: Challenges ahead," *Energy Policy*, vol. 38, no. 5, pp. 2291–2301, 2010, doi: 10.1016/j.enpol.2009.12.016.
2. M. Berglund, "Environmental systems analysis of biogas systems—Part I: Fuel-cycle emissions," vol. 30, pp. 469–485, 2006, doi: 10.1016/j.biombioe.2005.11.014.
3. K. Kaygusuz and J. Kirshner, "Energy services and energy poverty for sustainable rural development," doi: 10.1016/j.rser.2010.11.003.
4. M. Burguillo, "Assessing the impact of renewable energy deployment on local sustainability: Towards a theoretical framework," *Renew. Sustain. Energy Rev.*, vol. 12, pp. 1325–1344, 2008, doi: 10.1016/j.rser.2007.03.004.
5. S. Bouzarovski and Š. Robert, "Energy poverty policies in the EU: A critical perspective," *Energy Policy*, vol. 49, pp. 76–82, 2012.
6. M. El Fadel, G. Rachid, G. B. Boutros, and J. Hashisho, "Knowledge management mapping and gap analysis in renewable energy: Towards a sustainable framework in developing countries," *Renew. Sustain. Energy Rev.*, vol. 20, pp. 576–584, 2013, doi: 10.1016/j.rser.2012.11.071.
7. N. H. Afgan, "Sustainability assessment tool for the decision making in selection of energy system—Bosnian case," vol. 32, pp. 1979–1985, 2007, doi: 10.1016/j.energy.2007.02.006.
8. I. Dincer, "Renewable energy and sustainable development: A crucial review," *Renew. Sustain. Energy Rev.*, vol. 4, pp. 157–175, 2000.
9. Jänicke, M., "Green Growth: From a growing eco-industry to a sustainable economy (FFUReport 09-2011). Berlin. pp. 01–20, ISSN 1612-3026, Available: http://www.fu-berlin.de/ffu
10. J. R. S. Cristóbal, "Multi-criteria decision-making in the selection of a renewable energy project in spain: The Vikor method," *Renew. Energy*, vol. 36, no. 2, pp. 498–502, 2011, doi: 10.1016/j.renene.2010.07.031.
11. K. A. Khan, "Sustainable electricity generation at the coastal areas and the islands of Bangladesh using biomass resource." vol. 2, no. 1, pp. 09–13, 2016.
12. B. Sovacool, "The political economy of energy poverty: A review of key challenges Related papers." *Energy Sustain. Dev.*, vol. 16, no. 3, September 2012, pp. 272–282, doi: 10.1016/j.esd.2012.05.006
13. L. Gustavsson et al., "Using biomass for climate change mitigation and oil use reduction," *Energy Policy*, vol. 35, no. 11, pp. 5671–5691, 2007, doi: 10.1016/j.enpol.2007.05.023.
14. T. M. M. Verhallen and W. F. van Raaij, "Household behavior and the use of natural gas for home heating," *J. Consum. Res.*, vol. 8, no. 3, p. 253, 1981, doi: 10.1086/208862.
15. B. S. Reddy, H. Salk, K. Nathan, B. S. Reddy, H. Salk, and K. Nathan, "Energy in the development strategy of Indian households—The missing half energy in the development strategy of indian households—The missing half," *Renew. Sustain. Energy Rev.*, vol. 18, February 2013, pp. 203–210, doi: 10.1016/j.rser.2012.10.023
16. A. A. Afanas'ev, L. A. Bol'shov, and A. N. Karkhov, "Economic efficiency of new-generation nuclear power plants," *At. Energy*, vol. 81, no. 2, pp. 572–579, 1996.
17. J. Kim et al., "Development of advanced I & C in nuclear power plants: ADIOS and ASICS," *Nucl. Eng. Des.*, vol. 207, pp. 105–119, 2001.

18. S. C. Kaushik, V. S. Reddy, and S. K. Tyagi, "Energy and exergy analyses of thermal power plants: A review," *Renew. Sustain. Energy Rev.*, vol. 15, no. 4, pp. 1857–1872, 2011, doi: 10.1016/j.rser.2010.12.007.
19. F. Alobaid, N. Mertens, R. Starkloff, T. Lanz, C. Heinze, and B. Epple, "Progress in dynamic simulation of thermal power plants," *Prog. Energy Combust. Sci.*, vol. 59, pp. 79–162, 2017, doi: 10.1016/j.pecs.2016.11.001.
20. T. S. Wood and S. Baldwin, "Fuelwood and charcoal use in developing countries.," *Annu. Rev. Energy*, vol. 10, no. 9, pp. 407–429, 1985, doi: 10.1146/annurev.energy.10.1.407.
21. J. E. M. Arnold, G. Kohlin, R. Persson, and G. Shepherd, *Fuelwood revisited: what has changed in the last decade?*, vol. 37, 2003, doi: 10.17528/cifor/001197.
22. F. D. O. Falca, "Wave energy utilization: A review of the technologies," *Renew. Sustain. Energy Rev.*, vol. 14, pp. 899–918, 2010, doi: 10.1016/j.rser.2009.11.003.
23. K. A. Khan, S. Paul, A. Zobayer, and S. S. Hossain, "A study on solar photovoltaic conversion," *Int. J. Sci. Eng. Res.*, vol. 4, pp. 1–5, 2013, ISSN 2229-5518.
24. N. L. Panwar, S. C. Kaushik, and S. Kothari, "Role of renewable energy sources in environmental protection: A review," *Renew. Sustain. Energy Rev.*, vol. 15, no. 3, pp. 1513–1524, 2011, doi: 10.1016/j.rser.2010.11.037.
25. S. Mekhilef, R. Saidur, and M. Kamalisarvestani, "Effect of dust, humidity and air velocity on efficiency of photovoltaic cells," *Renew. Sustain. Energy Rev.*, vol. 16, no. 5, pp. 2920–2925, 2012, doi: 10.1016/j.rser.2012.02.012.
26. E. State and E. State, "Solar energy: A necessary investment in a developing," *Niger. J. Technol.*, vol. 23, no. 1, pp. 58–64, 2004.
27. S. C. Pryor and R. J. Barthelmie, "Climate change impacts on wind energy: A review," *Renew. Sustain. Energy Rev.*, vol. 14, pp. 430–437, 2010, doi: 10.1016/j.rser.2009.07.028.
28. K. A. Khan, S. M. Ahmed, M. Akhter, R. Alam, and M. Hossen, "Wave and tidal power generation wave and tidal power generation," *Int. J. Adv. Res. Innov. Ideas Educ.*, vol. 4, no. 6, pp. 71–82.
29. M. Berglund, "Environmental systems analysis of biogas systems—Part II: The environmental impact of replacing various reference systems," vol. 31, pp. 326–344, 2007, doi: 10.1016/j.biombioe.2007.01.004.
30. J. J. Chew and V. Doshi, "Recent advances in biomass pretreatment—Torrefaction fundamentals and technology," *Renew. Sustain. Energy Rev.*, vol. 15, no. 8, pp. 4212–4222, 2011, doi: 10.1016/j.rser.2011.09.017.
31. L. Mattila and T. Vanttola, "Strategy and research needs for nuclear power plant development: Plant modernization and possible new construction in Finland," *Nucl. Eng. Des.*, vol. 209, pp. 47–56, 2001.
32. S. Draria and A. H. Badi, "Controller for a nuclear research reactor," *Prog. Nucl. Energy*, vol. 46, no. 3, pp. 328–347, 2005.
33. W. Frisch and G. Gros, "Improving the safety of future nuclear fission power plants," *Fusion Eng. Des.*, vol. 57, pp. 83–93, 2001.
34. Y. Arata and Y. C. Zhang, "The basics of nuclear fusion reactor using solid pycnodeuterium," *Prog. Theor. Phys. Suppl.*, vol. 154, pp. 241–250, 2004.
35. L. Giusti, "A review of waste management practices and their impact on human health," *Waste Manag.*, vol. 29, no. 8, pp. 2227–2239, 2009, doi: 10.1016/j.wasman.2009.03.028.
36. D. Y. Goswami, S. Vijayaraghavan, S. Lu, and G. Tamm, "New and emerging developments in solar energy," *Sol. Energy*, vol. 76, no. 1–3, pp. 33–43, 2004, doi: 10.1016/S0038-092X(03)00103-8.
37. Y. A. Sadawarte et al., "Non conventional sources of energy," *International Conference on Emerging Frontiers Technology. Proceedings published in International Journal of Computer Applications (IJCA)*, pp. 1–11, 2012.

38. A. Mutalikdesai and S. K. Ramasesha, "Emerging solar technologies: Perovskite solar cell," *Resonance*, vol. 22, no. 11, pp. 1061–1083, 2017, doi: 10.1007/s12045-017-0571-1.
39. A. Hussain, S. M. Arif, and M. Aslam, "Emerging renewable and sustainable energy technologies: State of the art," *Renew. Sustain. Energy Rev.*, vol. 71, pp. 12–28, 2017, doi: 10.1016/j.rser.2016.12.033.
40. D. Barlev, R. Vidu, and P. Stroeve, "Innovation in concentrated solar power," *Sol. Energy Mater. Sol. Cells*, vol. 95, no. 10, pp. 2703–2725, 2011, doi: 10.1016/j.solmat.2011.05.020.
41. W. Ding and T. Bauer, "Progress in research and development of molten chloride salt technology for next generation concentrated solar power plants," *Engineering*, vol. 7, no. 3, pp. 334–347, 2021, doi: 10.1016/j.eng.2020.06.027.
42. G. Gaudiosi and C. R. Casaccia, "Offshore wind energy prospects," *Renew. Energy*, vol. 16, pp. 828–834, 1999.
43. GWEC, "Global wind report annual market update," *Wind Energy Technol.*, p. 76, 2015, Available: http://www.gwec.net/global-figures/wind-energy-global-status/.
44. J. Lee and F. Zhao, "Global wind report 2021," *Glob. Wind Energy Counc.*, p. 75, 2021, Available: http://www.gwec.net/global-figures/wind-energy-global-status/.
45. V. Yaramasu, B. Wu, P. C. Sen, S. Kouro, and M. Narimani, "High-power wind energy conversion systems: State-of-the-art and emerging technologies," *Proc. IEEE*, vol. 103, no. 5, pp. 740–788, 2015, doi: 10.1109/JPROC.2014.2378692.
46. O. Apata and D. T. O. Oyedokun, "Wind turbine generators: Conventional and emerging technologies," *Proc. – 2017 IEEE PES-IAS PowerAfrica Conf. Harnessing Energy, Inf. Commun. Technol. Afford. Electrif. Africa, PowerAfrica 2017*, pp. 606–611, 2017, doi: 10.1109/PowerAfrica.2017.7991295.
47. I. Kralova and J. Sjöblom, "Biofuels—Renewable energy sources: A review," *J. Dispers. Sci. Technol.*, vol. 31, no. 3, pp. 37–41, 2010, doi: 10.1080/01932690903119674.
48. T. G. Ambaye, M. Vaccari, A. Bonilla-Petriciolet, S. Prasad, E. D. van Hullebusch, and S. Rtimi, "Emerging technologies for biofuel production: A critical review on recent progress, challenges and perspectives," *J. Environ. Manage.*, vol. 290, p. 112627, 2021, doi: 10.1016/j.jenvman.2021.112627.
49. S. Hayashida, K. Ohta, P. Q. Flor, N. Nanri, and I. Miyahara, "High concentration-ethanol fermentation of raw ground corn," *Agric. Biol. Chem.*, vol. 46, no. 7, pp. 1947–1950, 1982, doi: 10.1271/bbb1961.46.1947.
50. R. A. Sheldon, "Enzymatic conversion of first- and second-generation sugars," Biomass and green chemistry, Springer, Cham, pp. 169–189, 2018.
51. S. Y. Lee et al., "Waste to bioenergy: A review on the recent conversion technologies," *BMC Energy*, vol. 1, no. 1, pp. 1–22, 2019, doi: 10.1186/s42500-019-0004-7.
52. R. Pelc and R. M. Fujita, "Renewable energy from the ocean," *Mar. Policy*, vol. 26, no. 6, pp. 471–479, 2002.
53. G. Nihous, "A preliminary investigation of the effect of Ocean Thermal Energy Conversion (OTEC) effluent discharge options on global OTEC resources," *J. Mar. Sci. Eng.*, vol. 6, no. 1, 2018, doi: 10.3390/jmse6010025.
54. J. Herrera, S. Sierra, and A. Ibeas, "Ocean thermal energy conversion and other uses of deep sea water: A review," *J. Mar. Sci. Eng.*, vol. 9, no. 4, 2021, doi: 10.3390/jmse9040356.
55. G. Jeerh, M. Zhang, and S. Tao, "Recent progress in ammonia fuel cells and their potential applications," *J. Mater. Chem. A*, vol. 9, no. 2, pp. 727–752, 2021, doi: 10.1039/d0ta08810b.
56. K. Jiao et al., "Designing the next generation of proton-exchange membrane fuel cells," *Nature*, vol. 595, no. 7867, pp. 361–369, 2021, doi: 10.1038/s41586-021-03482-7.

2 Digital Oil Fields and Its Emerging Technologies

Geetanjali Chauhan
Indian Institute of Petroleum & Energy, Visakhapatnam, India

Saurabh Mishra
Centre for Advanced Studies, Lucknow, India

Sugat Srivastava
Presidency University, Bengaluru, India

CONTENTS

DOI: 10.1201/b23013-2

2.1 INTRODUCTION

The petroleum industry has been a vital cog in the world's economic bloom by fulfilling the global energy needs in fueling, lighting, mobilization, and many more. In the years ahead, hydrocarbon energy consumption will continue to be the primary energy source, with a rising reliance on renewables contributing to more than 50% of the world's energy needs through 2040. Drilling a well in a subsurface rock formation, perforating the casing and transporting the energy to the surface via tubing in a controlled way through chokes and valves connecting to surface pipelines and separation and treatment facilities are all part of the oil and gas value chain. Trucks, pipelines, and ships convey treated crude oil and gas to refineries, where it is refined into refined products such as gasoline, diesel, and other fuels, and then ultimately delivered to consumers. The entire value chain operation necessitates several measures, which assist people involved in operation and production in monitoring various parameters such as flow rates, pressures, temperature, vibrations, and other factors to increase overall efficiency [1].

In an era, where all the industries are being empowered by digitalization and where digital connectivity in businesses has created value, the petroleum industry is also redefining its boundaries catching up the pace, overcoming the obstacles, and adopting new technologies related to the Internet of Things (IoT), Big Data, Artificial Intelligence (AI) and digital innovation. Table 2.1 lists the value that can be unlocked using digital solutions available in exploration, drilling, production, and field development domains of the upstream industry. The unprecedented challenges faced by the oil and gas industry, including shrinking size and fewer resources, uncertain and declining crude oil prices, declining production from conventional fields, geopolitics, concern about global warming, rising competition from other forms of energy, and globally distributed work teams, have led to the development and adoption of the digital technologies to extract and use hydrocarbon more efficiently and affordably from existing fields. The engineering is challenged to design cost-effective technologies to reduce the time to achieve maximum production with the greatest reliability and such objectives are sought to be fulfilled by DOF [2–6]. A typical schematic of DOF is shown in Figure 2.1.

Many people in the petroleum industry believe that digitalization began with SCADA (Supervisory Control and Data Acquisition) systems, which are analog systems with charts, and are manually read and interpreted. The DOF started on a small scale as costly technology in the mid-1990s with sensors, communication, and computation devices, followed by the advancement in digital technologies such as Big Data, AI, IoT, increased automation, collaboration decision centers, and focused work processes in the mid-2010s. The digital developments in the industry followed a slow growth for decades, working mostly on the monitoring aspects mainly because developments in multiphase flow meters, sensors, high-pressure–high-temperature

TABLE 2.1
Value Unlock in Upstream Activities Using Digital Solutions [2]

Exploration	Field Development	Drilling	Operations/Production
Around 60% Decrease in the time and cost for data interpretation	Around 70% Decrease in engineering and field development-related time	20–30% Quicker well delivery and more productive well	3–5% Increased production 20–40% Reduced maintenance Cost
		Digital Solutions	
– Accelerate interpretation with Machine Learning	– Optimize field architecture with integrated modeling – Synchronize project build using digital twins	– Closed-Loop Automation for faster well delivery – Improved well design using data analysis	– Optimize production with real-time data and advanced models – Improve uptime using predictive maintenance

FIGURE 2.1 Typical schematic of DOF.

gauges, and high-frequency computational software happened recently [6]. However, still today the DOF is not fully leveraging the breakthroughs in communication and information technologies in real-time field management such as IoT, AI, and Big Data and the golden era in DOF is yet to come in coming decades [1].

Table 2.2 highlights the four business operation levels of the oil and gas industry, starting with manual operations and progressing through automation, real-time operation center (RTOC), and ultimately DOF which reflects the enhancement in the performance in terms of production uptime and operational efficiency [1].

TABLE 2.2
Business Operation Levels for Oil and Gas Industry [1]

O&G Business Operational Levels	Infrastructure	Advantages
Manual (Level 1)	– Basic surface sensors – Pressure (P), Temperature (T) – Reading recorded by person on physical visit of site – Data Shared by mail or shared repository	– Low assets collaboration
Automation (Level 2)	– Basic SCADA sensors – P, T – Well test data monthly – Daily monitoring – Diagnostics and well actions monthly – Nonintegrated data transfer – Some workflows automated	– Better collaboration – OpEx reduction
Real-Time Operation Centre (RTOC) (Level 3)	– SCADA sensors – P, T – Virtual metering, real-time monitoring – Diagnostic and operations actions weekly – Some automated workflows – Mostly manual controls/intervention – Most workflows automated – Structured data environment	– Production increase – OpEx reduction – Downtime management
DOF (Level 4)	– Multiple P, T, Vol, downhole real-time data (RT) – Diagnostic and optimization in RT – Intelligent and fit-for-purpose workflows – Automated controls – Integrated data systems	– Improved production – Downtime mitigation – Optimized performance

The lowest level, i.e., manual which involves no automation, is executed completely by humans with minimal collaboration. Depending on the size and age of the assets, the operations may operate at several levels simultaneously. For example, mature field operations can be completely manual to greenfield operation working on SCADA or semi-automated with some sensors, to fully automated with many sensors. Table 2.2 also lists the infrastructure used in each level along with the benefits obtained.

The DOF includes present-day cutting-edge technology utilizing flow meters, sensors, telemetry channels, and combining it with data analytics, for efficient decision-making, thus maximizing the production of hydrocarbon by delivering the right data to the right team at the right time for effective decision-making and improving operational efficiency. RTOC is the integration of real-time sensors, SCADA system, wireless data communication, automated systems, and a dedicated real-time operation center. DOF is an advanced adaptation of RTOC with the same operations but with intelligent workflows, predictive capability involving specialized collaboration working environment (CWE) with dedicated staff, and workflow mobile communicating via video, closed-circuit television as well as chatting with field staff.

Real-time monitoring reduces production downtime through day-to-day diagnostics and optimization integrated with an advisory system [1].

The DOF is built by integrating advanced technologies in automation, data communication, and computing with work processes and skilled people, followed by thorough analysis to sustainably minimize OPEX & CAPEX and environmental impact while getting maximum out of the field in terms of the production. It also ensures the integrity of linked equipment and enhances the safety of the crew involved [4–5].

The solution driven by this software could prevent E&P company's digital field project from becoming a SCADA project which means that after instrumenting measurement and SCADA/DCS system in the field, there are only surveillance/monitoring functions for the field with very few analysis perspectives. This is also a big waste.

2.1.1 DOF Cycles and Current Technologies

The vital objectives of DOF (Figure 2.2) based on the data gathered from the field can be categorized into three domains: (1) proactive production analysis, (2) realistic production optimization, and (3) detailed reservoir management. The DOF cycle domain referred to as *Operations*, which requires a decision time up to 1 day, is supported by in-place SCADA by acquiring data from diverse areas such as flow line sensors, permanent subsurface well, and surface facility instrumentation to manage control valves for real-time operational activities including controlling remote equipment and failure detection [6].

With a decision period of within three months, the *production analysis/optimization decision cycle* integrates field production data to observe and analyze current production system behavior to optimize well drawdown to restrict water/gas coning, control sand, and maximize production. The *field management decision cycle* is the slowest and takes months to years for decision and entails planning reservoir depletion strategy, deciding the location of in-fill wells based on the integration of field

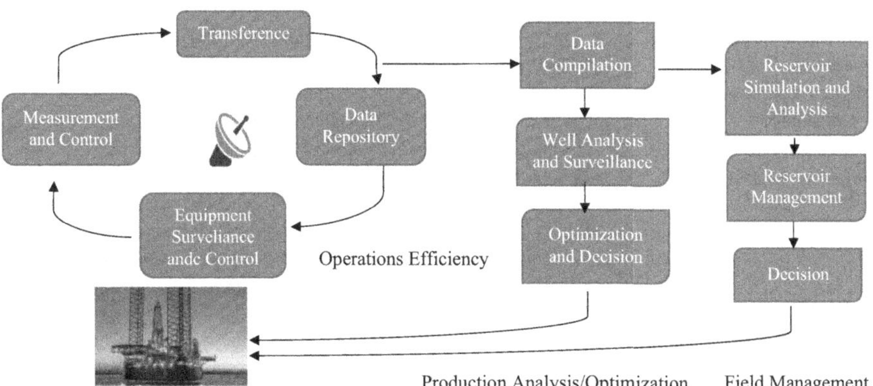

FIGURE 2.2 Digital field areas and their life cycles.

data as well as construction and simulating reservoir model to plan the field development scenarios considering the economics and other parameters [6].

DOF entails the combination of technology-centric solutions to maximize the use of limited resources. DOF has become ideal for the development of green fields and is slowly taking over the brownfields. DOF is no longer an option for the companies; it has evolved into a critical function. Be it faster and reliable seismic interpretations using AI, to remotely monitor the locations using drones, robots; digital is deluging many solutions to oil industry vulnerability issues like oil price crashes and unlocking symbolic values in all the areas of the industry. The following are a few technologies (as shown in Table 2.3) that have gained widespread popularity in DOF and are allowing real-time or near-real-time analysis to enhance well, reservoir, facility, and asset performance.

TABLE 2.3
Technologies Used in DOF and Their Key Benefits [3, 7]

Technology	Key Benefits
Remote Real-Time Facility Monitoring and Control	SCADA networks enable transferring offsite data to control rooms for monitoring and control of valves and pumps in real-time.
Real-Time Drilling	Allows collection of real-time data such as Rotation Per Minute (RPM), Weight on Bit (WOB), downhole pressures to effectively analyze ongoing drilling and controlling it via remotely steerable down-hole tools, anti-collision optimization
Real-Time Production Surveillance	The employment of advanced alarm systems to maintain the installed production capacity levels, smart well monitoring and control, artificial lift diagnosis and optimization, real-time integrated production optimization, virtual metering: multiphase flow rates, soft sensors, and zonal allocation.
Intelligent Wells	The employment of surface-controlled equipment for down-hole fiber optic sensors for continuous monitoring of conditions and response.
4-D Visualization and Modeling	Additional insight into production enhancement via successive 3-D seismic surveys that track fluid movements for enhanced recovery mechanisms.
Workflow and Knowledge management Systems	Robust document-management and historical data solutions to quickly execute workflows and routines.
Remote Communications Technology	Control room facilities with real-time visual, voice, and data communication with the field allowing more quick and analytical responses by a mix of field and office staff.
Integrated Asset Models	Applications that model complete production system performance from the producing horizon, through the well-bore, through the production facility, and onto the export/sales point across disparate data sources and multisite work teams. Includes automated reservoir history match, smart water flooding, integrated planning and scheduling, twin modeling of the subsurface.
Reservoir Management	A cost-effective way to develop a new field or to bring new life to a mature field with enhanced oil recovery measures such as steam flooding or steam injection to improve production rates and increase the total amount of oil and gas recovered from a field.

2.2 COMPONENTS OF DOF

To implement DOF solutions, various sensors and actuators (micro/nano), integrated circuits, logical algorithms, and computers are coupled with specialized digital instrumentation units that are fully synchronized with a crew, instrument, and work process [1]. DOF components are classified as follows:

1. Digital instrumentation
2. Process automation and control
3. Data management and transmission
4. User interface and visualization
5. Collaboration and people organization

2.2.1 Digital Instrumentation

An oil field encompasses several processes and equipment on the surface and downhole. Specially designed micro-electromechanical system (MEMS) devices, which can work under high-temperature and high-pressure (HTHP) conditions, are placed in downhole locations to monitor the reservoir pressure, reservoir temperature, and flow of reservoir fluid. These MEMS devices comprise sensors that transmit the data through an electrical cable to a control panel on the surface to enhance the reservoir understanding resulting in optimized production and improved recovery factor.

Subsurface instrumentation coupled with subsurface control refers to smart completions. Inflow control valves (ICVs) and inflow control devices (ICD) are in practice to actively control the flow of reservoir fluid from the different formation zones and to maintain the overall wellbore stability. The ICVs are pre-installed valves with the completion to control the reservoir fluid flow from the target zone which are controlled and actuated electrically through a telemetry system on the surface and limited by the available wellhead and packers control line penetration [8].

ICDs are desirable to optimize the production from the targeted zone and control the unwanted reservoir fluid flow into the wellbore [9]. ICD systems stabilize the production profile by creating a pressure drop and thus balancing the drawdown from the different production zones. This function of the ICD system reduces the probability of water/gas conning, especially in horizontal wells.

Downhole monitoring systems or gauges (i.e., electrical sensors, detectors) are essential for monitoring downhole conditions and transmitting real-time data from the wellbore to the surface. Thus, with the assistance of these monitoring systems, any changes in wellbore conditions or reservoir fluid properties can be observed at the surface within seconds for the necessary action. These are useful for noting the performance of the well under all normal and critical conditions. The downhole gauges can be installed anywhere in the well to meet the operator's objective and collect the data regarding temperature, flow rate, and composition of the reservoir fluid. The changes monitored by the sensors and detectors are received by the telemetry systems via the electrical wireline, as indicated in Figure 2.3.

The process facilities on the surface, oil refineries, petrochemical industry, pipelines, and distribution units also utilize digital instrumentation to regulate and

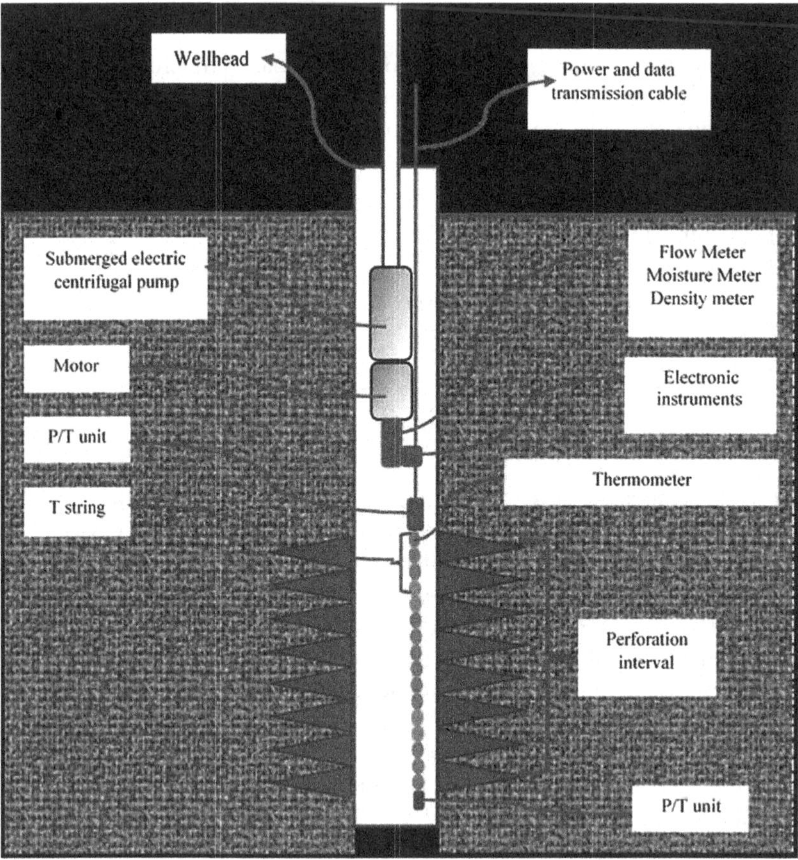

FIGURE 2.3 Wellbore monitoring system [10].

monitor the physical, chemical, and biological phenomenon that occur at different conditions. The sensors in the system operate as an eye, nose, and ear for the environment, sensing phenomena and sending the collected data to an electronic device, which processes it like a brain to make judgments. This digital equipment aids in the control of emissions/pollution from various stages of processing in the industry, as well as alerting nearby personnel to potentially hazardous situations.

2.2.2 Process Automation and Control

Personnel working in the petroleum sector, both upstream and downstream, are involved in a variety of processes and operations, dealing with a large amount of data at various stages of the process. The manual process entails a large amount of manual engagement with operational data as well as its analysis. This could result in time wastage and miscommunication, leading to fewer efficient procedures and a higher risk of errors. Oil field automation involves the development and application of technology to monitor and control exploration, field development, production, and

TABLE 2.4
Manual Workflow vs Automated Workflow

Manual workflow	Automated workflow
Decentralized information source	Standardized and controlled process
Repetitive tasks	24/7 reliability and consistent quality of process execution
Low productivity	Central storage for efficient document management
Manual communication channels	Accurate information and consistency in data output
Human error and inaccurate data	Minimize input error
Lost processing time	Automatic and transparent communication channels
Time-consuming for engineers and managers	Sharing data at the same time

petroleum products' delivery. Automation comprises a wide spectrum of technologies touching the domains of robotics and expert systems, telemetry and communications, electro-optics, cybersecurity, process measurement and control, sensors, wireless applications, system integrations, and many more. It helps the oil industry, determined to keep the cost down by imitating users in carrying out routine tasks with speed and accuracy. Automation not only reduces the time required for the operation, but also helps the engineers and operators to have rapid access to crucial system performance, conditions, and technical information which improves the efficiency of performance and decision-making. Table 2.4 shows a comparative study between manual and automated workflow in the oil industry.

2.2.3 Data Management and Transmission

The petroleum industry involves a plethora of operations. Many times, these operations deliver a massive volume of unstructured data in the form of documents, images, maps, and spreadsheets, referred to as "Big Data" and the industry face challenges and limitations in processing these data due to: (1) inadequate usage of trained manpower in non-substantive activities involving data search and collection; (2) redundancy in tasks by operational team separately; (3) unawareness of the existence of such information/data, culminating in the loss of valued business visions; (4) sluggish processing and delay in output, affecting overall decision-making in operations; (5) data are available in file formats difficult to extract; (6) loss in stored information/data in knowledge transfer between old and new employees.

To overcome, the limitations and challenges in extracting/managing data, the petroleum industry uses a digital platform to manage and deploy the complex technical data in usable form. The digital platform can be segmented into three categories: (1) IT infrastructure; (2) data management and transmission; (3) services, and these play a vital role in achieving high operational efficiency. The oil and gas industry collaborates with the data expert to get the right data to the right place at right time. These data experts extract the deep quality insights of any data type from any source. To execute this action, they need full data traceability, full access to data, and complete version history. The data flow on the digital platform and users can always access it at the individual/group level and job-based level. Moreover, the data remain encrypted and safe even in transit conditions. There are so many market players such as SAP, IBM, Wipro, NetApp, and Oracle, who provide on-premise and cloud-based data management to oil and gas sectors. These vendors utilize AI technologies such as Optical Character Recognition (OCR), Natural Languages Processing (NPL), and Knowledge Graph (KG) to extract and connect the information from different data sources. Recent advances in machine learning (ML) and IoT for data search applications made it efficient to extract structured information from the free texts found in observer's log/well commentaries, reservoir reports, datasheets and logs.

2.2.4 User Interface and Visualization

In the oil and gas sector, visualization is a powerful tool for representing/understanding a set of data or a surface/subsurface event. In the context of data and information, the oil and gas industry and others posed a question like "can a machine be given a human element that can consume/analyze all the data generated and predict the event just like a human does". Digitalization is the answer to the question. The software learns to emulate human intelligence based on the actions of subject matter experts. This machine learns to study and anticipate future event occurrences by analyzing the generated information/data at various zones of operation. Furthermore, the user interface offers an advanced visualization tool as well as a user-friendly approach to communicate with different operational zones. This user interface also facilitates team communication and manages information management challenges such as user access and data security in a transparent manner. The visualization facilitates communication between personnel from many disciplines such as geology, geophysics, and petrophysics and allows them to speed up their analyses to extract essential insights from the data/information.

2.2.5 Collaboration and People Organization

Traditionally, petroleum industry has worked in silos; however, a successful DOF requires working in collaborations among all the departments working in a value chain. A collaborative working environment allows people of the organization with varied fields of knowledge to provide qualitatively good inputs leading to an efficient and successful project outcome. Collaborative work is built on the idea of linking offices, people, and fields for a single asset in an organized procedure on a digital platform. Men, machines, and materials are all monitored as well as maintained, and high-definition video communication is set up on several screens to share important information. It is well acquainted that telephone or video conferences do not meet the varied requirements of collaboration between remote locations. The virtual room concept can provide the solution to these diverse requirements [11]. This collaboration and organizing personals together result in optimized operations with real precision. Figure 2.4 shows the virtual link in a collaborative work environment between on-site, offsite, and Central facilities. By working together, users may deploy, discuss, analyze and monitor information, and make decisions.

FIGURE 2.4 Virtual collaboration between personnel from the remote location and the central facility.

2.3 PROVED DIGITAL SOLUTIONS FOR PETROLEUM INDUSTRY

The key objective of the DOF is to maximize the recovery factor, eliminate non-productive time, and increase the economic return and deployment of automated workflow. Table 2.5 shows few examples of real-world applications of DOF initiatives by various organizations and the key benefits observed.

2.4 CHALLENGES IN IMPLEMENTING DOF

The DOF technology implementation carries vivid changes in the management process. In DOF initiatives, success is contingent on successfully managing the organization's and people's complexity. The DOF has been in use for more than two decades now; however, the petroleum industry still faces many obstacles to achieve the anticipated potential from Digital technologies mainly on two fronts. First, blending suitable technology into the existing oil fields, including Information Technology, and second, overcoming the initial ripples in terms of efforts required for the acceptance of digital technologies by the organizations and its people as well as overcoming its challenges like resources and funding in unrevealing the value of digitalization [7]. The factors that are preventing DOFs from being fully implemented are discussed in the section below.

2.4.1 Culture and Mindset

Digitalizing the oil fields have a direct impact on the organizations at the executive and field levels requiring personnel to upgrade with the new skills and employing competent specialists, making senior management apprehensive due to the conservative nature and the possible consequences of such changes. This conservative nature results in inadequate encouragement from management, failing in compelling for the

TABLE 2.5
Digitalization in Different Oil Fields [12–20]

Operator Company	Field(s) for Digitalization	Operational Zone of Digitalization	Technologies Deployed	Key Benefits Observed
Linn Energy-Berry Petroleum	San-Joaquin Valley, Mid-Continent Gas, Permian, and Uinta Basin	Oil production, steam injection, gas management,	CWE, Real-time monitoring, Data integration, Mobile communication	– Increase in production and steam efficiency. – Decrease in well failures. – Reduced data analysis time. – Faster decisions.
Kuwait Oil Company	Kuwait Integrated Digital Fields Project	Production Operations, Artificial lift, Water encroachment	Automation, real-time monitoring, ESP coupled with MEMS devices	By increasing water injection quantity by 30 MSTB in the region containing 60 wells: – oil output increased by 8% per day, In just ten proactive steps per well, they were able to raise overall oil production by 37%. – Enhanced team efficiency with a reduction in the time needed to study one well from 7.3 to just 1.6 hours.
Diamondback Energy	Permian Basin Fields	Completion optimization in Unconventional resources	Machine Learning (ML)	– Delivered reduced completion cost by 5 to 32% per well. – maintaining similar production rates.
Petrobras	Brazilian Fields	Production operations, workover	Real-time monitoring, Automated workflow, CWE, Digital Instrumentation, Data analytics	– Increased production and operational efficiency. – increased recovery factors.
Shell E&P	Mark I Nelson Field, North Sea Fields	Business Value	Real-time well monitoring and optimization, data acquisition and control, architecture security, CWE	– Efficient well optimization and reservoir optimization achieved a significant profit of 5 billion dollars. – Oil production increased by 15% due to the optimization of gas lift operation. – Automatic data transfer resulted in increased efficiency by 2 hours/day. – Reduction in manual data by 4 hours/day. – Operational reliability increased by several percent.

technology adoption and not taking a risk by experimenting and "fail first" approach. The traditional work practices which include working in siloed environments need to be broken for effective communication and interchanging roles across diverse operating groups. At field levels, lack of willingness to learn new skills, preferences of the personnel toward manual workarounds, and not trusting the technology, is an issue. Organizations must effectively manage changes in terms of roles definition and decision rights at the site and offsite, as well as aligning service company employees into the new interlocking model [3, 7].

2.4.2 Skill Gap Is Real

The best technology can fail if the user lacks the necessary abilities to comprehend the technology's investment worth and its use for increasing efficiency. The oil and gas business is facing disruptions due to a skills shortage that prevents it from fully leveraging from digital. There is a high level of immaturity across the industry in terms of the skills needed to extract profit from digital technologies such as AI, data analytics, data science, cyber security, design thinking, and others. There is also a need to reskill and upskill 60% of the existing oil and gas industry workforce to justify and gains from the investments being done into digitalization. Additionally, pulling in digital skills personals is also a challenge nowadays as the young generation has more inclination toward alternative greener industries.

2.4.3 Ecosystem

The oil industry traditionally has not been very good at sharing data even with an organization. Some businesses continue to use the traditional methods of storing data on papers and spreadsheets combined with human proficiency in taking important decisions. In the era of digitalization, the Data security regulations no longer fit the purpose of adding value within an organization. The data sharing or integration from equipment, systems, and sensors across the industry's value chain is yet to be adopted where the organizations should feel secure that by sharing the data they are going to be more efficient in terms of asset management, productivity, and safety [7].

2.4.4 Lack of Standardization

The industry still needs to standardize or integrate sensor data in terms of data ownership and quality issues such as large volumes, incomplete datasets, multiple file formats, and the complex nature of data in inconsistent formats, which necessitates a significant amount of time for cleansing before proceeding with the analysis [7].

2.4.5 Cybersecurity

In the chase to expand in digital technology to include automated infrastructure, connected computing devices, servers and applications, web-based telecommunications systems, and the totality of data storage and transmission systems, the biggest risk is of cyberattacks led by hackers and terrorists, where crucial industry like oil and gas will have to lose much more than just reputation.

The industry has witnessed many security breaches, primarily at offshore sites, including hacking of industrial control systems, cyber espionage, malware-infected programs, hydrocarbon installation terrorism, undetected spills resulting in plant shutdown/sabotage, and production disruptions, which has resulted in the loss of Company's assets, production hampering and significant financial losses. The oil and gas companies cannot afford to be vulnerable to cyber-attacks and hence need to spend more on essential defensive countermeasures and employee training to reduce human error [7, 21].

2.5 FUTURE POTENTIAL AND EMERGING TECHNOLOGIES

2.5.1 Future Market Growth

The oil prices are volatile and suffer from periodic extremes in the prices. Also, the industry itself requires high initial investments with break-even points occurring after 4–5 years of constant production that are also marred with a plethora of uncertainties, risks, personnel working in remote and dangerous conditions, and sometimes, encountering a dry (or not so profitable) well/s. Thus, since its inception, it takes more than ten years to reach the break-even point. To survive in this volatile and risky market, companies are finding ways to reduce their expenditure. In recent times, switching to DOF technology has proven to be one of the most cost-efficient methods [3]. The production spending between 2014–16 dropped by 29% as a precautionary measure [22], owing to risk aversion, and the cost-cutting attitude of the industry. The industry now hopes that the recent advancements in the DOF sector might prove to be the game-changer. According to the white paper by World Economic Forum [23], DOF technologies are expected to save around [22]:

1. 20% in drilling and completion
2. 25% in inspection and maintenance
3. 20% lower employee costs

According to the report compiled by Market Watch, the global market of the DOF is projected to reach $ 35 billion by 2030, with a healthy CAGR of 8.5% [24]. Another report by Market Research Future (MRFR) prescient that the global oil field market will hit $ 28.61 billion at a CAGR of 6.5% between 2020 and 2027 [25]. Although both these reports predict different values of market cap and CAGR, one thing is certain that the market of DOF is going to boom in the coming future. The catalyst for this trend is the advent of new technologies, which emphasize remote control of oil fields for process optimization and automation to optimize production and reduce unproductive time.

Another reason for the projected boom in the market is the growth of the middle class' life standard. Around the world, the middle class is expected to more than double by 2035 [25]. Due to this effect, the energy consumption will surely increase with an increase in the sale of vehicles, buying the latest electronic goods, etc.

There is a huge disparity in the market share when different regions are compared. In 2018, Europe became the biggest DOF development market, accounting for almost 30% of its sale [24]. The biggest contributor to such a trend is the increase in the demand for digital technology and its acceptance by Oil and Gas companies. Among the European nations, the United Kingdom had the biggest share of more than 25%, in terms of volume; primarily due to the existence of huge North Sea reserves. These fields are matured with a decline in the production rate, thus pushing the upstream companies to improve the recovery rates with the help of DOF technology. In terms of revenue generation using DOF technology, Saudi Arabia tops the list in 2018.

2.5.2 Future Potential with Emerging Technologies

Assimilation of technologies is a dynamic process and it keeps evolving with the advent of newer technologies. The amalgamation of the oil and gas industry and recently evolved technologies should pass the test of time to have credibility for investing in such an extravagant money affair. Thus, extrapolating the future for any technology and integrating it with the oil and gas industry is next to impossible. The following are some of the major technologies which can interrupt the current oil field scenario in the next 5–10 years: (i) Ubiquitous Sensor Networks (USN) and Industrial Internet of Things (IIoT); (ii) Big-Data Analytics; (iii) AI; (iv) Mobility and Extended Reality.

The current scenario focused mainly on automation, and optimization of workflow to improve the safety of manpower and work-rate of the operation. The next phase will surely aim to go deeper into these facets by incorporating the recent developments in silicon technology and Nano Electro-Mechanical Systems (NEMS). The sensor technology will bring automation and accuracy in data acquisition, whereas Big-Data analytics will work on finding the trends to make decisions based on the collected data. To make the analysis faster and more accurate, an AI-based algorithm can be used and then incorporated into the technologies like Reinforcement Learning, Deep Learning, etc. In the future, usage of phones and tablets, and Extended Reality technologies will affect field management.

2.5.2.1 Ubiquitous Sensor Networks and Industrial Internet of Things

The IIoT refers to a network of interconnected sensors, instruments, data, processes, and other devices which communicate with each other seamlessly and enhance industrial operations. These candidates will collect the data/information, interact with the surrounding, and communicate via the internet. The advent of 5G connectivity will catalyze and smoothen the transmission of the information at the digital platform. The devices, connected by a high-speed 5G network, will collect the data, work on the field environment, and transmit over the internet. For implementing these arduous tasks, sensors are required "ubiquitously"-sensors that can interact with the surroundings and processes throughout time and space.

Ubiquitous sensors are the next generation of Wireless Sensor Area Network (WSAN) [26]. The sensors are connected to add scalability through different mesh networks. The future objective of USN can be fulfilled by miniaturizing these sensors to a nanoscale so that they can be used in remote places inside the reservoir, where conventional sensors do not work. In this way, nano-sensors can provide real-time key insights of the reservoir variables like reservoir pressure, temperature, WOC, GOC, etc., resulting in an open platform to IIoT to monitor the operations efficiently.

The IIoT system monitors variables such as pressure, gas flow, compressor conditions, temperature, concentration, etc., to detect leakage thus saving millions of dollars as well as neutralizing the environmental risks. The smart PIG (Pipeline Inspection Gauge), which remains inside the pipe, detects cracks of the sides and welding effects using ultrasound waves and magnetic flux features [27]. For gas leak detection, the system uses a camera with a filter that is sensitive to a selected range of infrared wavelengths.

2.5.2.2 Big-Data Analytics

Big-Data Analytics can manage huge data sets accurately and has attracted the petroleum industry for its robustness in information/data collection and processing as well as in analysis for efficient exploration and production operations. Big-Data Analytics is the usage of advanced analytical techniques on huge, diverse data sets. These data sets can be structured, unstructured, and multi-structured and can include various data formats due to multiple combinations of people/machine interactions. The data size may vary from terabytes (10^3 GB) to zettabytes (10^{12} GB). A typical Big Data analytics system consists of layers, i.e., data acquisition and monitoring layer, the transmission layer, and the data analysis and production management layer; with each having a specific task. Data acquisition and monitoring can be attained by using actuators and sensors [28]. The transmission layer is like the "pipeline" for data transfer and it "transports" all the data collected by the previous layer through a secured channel. The final layer combines and analyzes all the received data (can also perform data visualization) which can be used to prepare a business model.

To improve an organization's business processes, methodologies like Six Sigma, Lean Six Sigma, Kaizen, etc., are becoming common. Six Sigma aims at improving the performance and decrease in process variation, which culminates in greater profits, efficiency and quality of products [29], through statistical analysis.

2.5.2.3 Artificial Intelligence

AI systems are well recognized to automate and optimize the initial stage of the exploration and production lifecycle. In general, AI is defined as the simulation of human intelligence processed by machines [30]. The market player who adopted it early is clinching a competitive advantage in protecting their assets as AI technologies are proving their worth in optimizing operational activities. AI systems have shown significant outcomes in reducing risks, enhancing production, and minimizing operational costs. The demand for AI is already growing in the petroleum industry as the utilization of advanced digital technologies is increasing. AI-based optimization methods (optimization methods in general) are used for determining different operational parameters like asphaltenes precipitation, minimum miscibility pressure (MMP) well placement, history matching, drilling operation, pipeline condition, etc. Using conventional optimization techniques like linear, discrete, integer, etc., can quicken the process but each of these has its limitation which is based on the nature of the objective function. To generalize and faster the calculation, non-linear programming techniques like AI-based methods (mainly comprised of Evolutionary Algorithms (EA), Swarm Intelligence (SI), Fuzzy Logic (FL), and Artificial Neural Networks (ANN)) became prominent and are used more often in recent times [31].

2.5.2.4 Mobility, Collaboration, Virtual, and Augmented Reality

By creating mobile or tablet applications (apps) for the field data and dashboard, personal can remain connected to the field operation even on the move. Using such an application, decision-making can become faster and easier. Moving a step further and integrating wearables can further improve mobility. One of the applications of mobility and collaboration was reported by Ailworth, 2017 [32] which talks about using an EOG resources (an American energy company) app by rig personal to change the direction of bits with real-time data and real-time communication with central offices. Following a similar trend, other companies are developing and utilizing similar types of apps to field personnel, pumpers, instrument techniques, etc., in their company.

Extended reality (XR) technology had rapidly transformed the gaming and entertainment industries and is set to change the oil and gas industry as well. Extended reality (XR) is the umbrella term that includes technologies like Virtual Reality (VR), Augmented Reality (AR), and Mixed Reality (MR). These technologies enhance our senses by providing additional information about the actual world and or creating a simulated unreal world for us to experience. Augmented Reality (AR) adds digital elements to a live view often by using the camera on a smartphone. Virtual Reality (VR) implies a complete immersion experience that shuts out the physical world. By wearing a VR device on the eyes, users can experience situations like riding on the back of a dragon, observing the environment in distant Antarctica, etc. Mixed Reality (MR) is a step beyond AR and combines elements of both AR and VR. It adds additional information for a user to perceive, in MR, the user can interact with the interesting physical and virtual worlds, thus blurring the difference between real and virtual.

In the coming future, XR will become more prominent in the oil and gas industry. VR can be used to create an immersive 3D environment with 360-degree views to add more detail to a process. Shell, Chevron, and Exxon-Mobil are using VR for training purposes, which reduces the training costs as well as increases the depth of training by simulating the environment [1]. Augmented Reality (AR) can be used for the dynamic interaction of field equipment by trainees and can be operated remotely without moving, travel, or waiting for instructors. The integration of XR technologies with the oil and gas industry is going to increase with time. Soon, the companies might use VR for training purposes and AR for operating multiple fields by a single operator more often. AR and VR can also be utilized to understand the sub-surface environment for geologists and reservoir engineers for decision-making and improving the existing models.

2.6 CONCLUSIONS

DOF is a revolutionary technological solution that has the capacity of bringing an indispensably powerful organizational model—leveraging technical and cross-functional expertise for better decision-making and enhancing productivity. Its applications like real-time monitoring and CWE align multiple diverse units into a unified/shared operating unit to work cohesively and allow multilocation and multifunctional teams to work efficiently breaking the conventional environment of working in silos. However, transiting into DOF employment is not that smooth and requires funds allotment, process change, management change, and up-gradation of new skills which creates initial ripples in the organization for its acceptance. The inclusion of the latest technologies such as Cloud, Big Data, AI, and IIOT in the DOF will provide a powerful environment with extensive end-to-end automated workflows to help in building intelligent solutions which can self-learn, monitor, and control the equipment and hardware in the field. Hence, the organizations need to evaluate their strategy to become more holistic in addressing key challenges in implementing DOF.

REFERENCES

1. Carvajal, G., Maucec, M., Cullick, S., The Future Digital Oil Field (Chapter Nine), Editor(s): Carvajal, G., Maucec, M., Cullick, S., *Intelligent Digital Oil and Gas Fields*, 2018, Gulf Professional Publishing, Pages 321–350, ISBN 9780128046425. https://doi.org/10.1016/B978-0-12-804642-5.00009-8
2. Vision Paper, *Digitalization Roadmap for Indian Exploration and Production (E&P) Industry*, June 2021
3. Saputelli, L., Cesar, B., Michael, N., Carlos, L., Cramer, R., Toshi, M., Giuseppe, M., Best Practices and Lessons Learned After 10 Years of Digital Oilfield (DOF) Implementations, SPE-167269-MS, 2013, Presented at the *SPE Kuwait Oil and Gas Show and Conference*, Kuwait City, Kuwait, October 2013. https://doi.org/10.2118/167269-MS
4. Al Qahtani, A.M., Al Qahtani, M., Adding More Value in the Downturn Time from Digital Oil Field; What Is More to Leverage?, SPE-192294-MS, 2018, Presented at the *SPE Kingdom of Saudi Arabia Annual Technical Symposium and Exhibition held in Dammam*, Dammam, Saudi Arebia, 2018. https://doi.org/10.2118/192294-MS
5. Cramer, R., Göbel, D., Mueller, K., Tulalian, R., A Measure of Digital Oil Field Status–Is It The End of The Beginning?, SPE-149957-MS, 2012, Presented at the *SPE Intelligent Energy International held in Utrecht*, Utrecht, The Netherlands, 2012. https://doi.org/10.2118/149957-MS
6. Al-Jasmi, A., Qiu, F., Ali, Z., Digital Oil Field Experience: An Overview and a Case Study, SPE-163718-MS, 2013, Presented at *SPE Digital Energy Conference and Exhibition held in The Woodlands*, The Woodlands, TX, 2013. https://doi.org/10.2118/163718-MS
7. Steinhubl, A., Klimchuk, G., Click, C., Morawski, P., *Unleashing Productivity: The Digital Oil Field*, 2008, Advantage, Booz & Company.
8. Wang, J., Zhang, N., Wang, Y., Zhang, B., Wang, Y., Liu, T., Development of a Downhole Incharge Inflow Control Valve in Intelligent Wells, 2016, *Journal of Natural Gas Science and Engineering*, Vol. 29, pp 559–569. https://doi.org/10.1016/j.jngse.2016.01.020.
9. Li, Z., Fernandes, P., Zhu, D., Understanding the Role of Inflow-Control Devices in Optimizing Horizontal-Well Performance, 2011, *SPE Drilling & Completions*, Vol. 26(3), pp 376–385. https://doi.org/10.2118/124677-PA
10. Ulyanov, V.N., Cheremisin, A.N., Toropeckij, K.V., Ryazantecv, A.E., *Downhole Monitoring of Smart Wells*, 2016, ROGTEC Russian Oil and Gas Technologies. https://rogtecmagazine.com/downhole-monitoring-smart-wells/
11. Gulbrandsoy, K., Hepso, V., Skavhaug A., Virtual Collaboration in Oil and Gas Organizations, 2002, *ACM SIGGROUP Bulletin*, Vol. 23(3), pp 42–47. http://doi.org/10.1145/990017.990026

12. Betz, J., Data Integration Enables Quicker Decisions, 2015, *Journal of Petroleum Technology*, Vol. 67(5), pp 76–77. https://doi.org/10.2118/0515-0076-JPT
13. Eldred, F., Cullick, A.S., Purwar, S., Arcot, S., Lenzsch, C., A Small Operator's Implementation of a Digital Oil-Field Initiative, SPE-173404-MS, 2015, Presented at *SPE Digital Energy Conference and Exhibition*, The Woodlands, TX, 2015. https://doi.org/SPE-173404-MS
14. Dashti, Q., Al-Jasmi, A.K., AlQaoud, B., Ali, Z., Bonilla, J.C.G., Digital Oilfield Implementation in High Pressure and High Temperature Sour Environments: KOC Challenges and Guidelines, SPE-149758-MS, 2012, Presented as *SPE Intelligent Energy International*, Utrecht, The Netherlands, 2012. https://doi.org/10.2118/149758-MS
15. Al-Abbasi, A., Al-Jasmi, A., Nasr, H., Carvajal, G., Vanish, D., Wang, F., Cullick, A.S., Md Adnan, F., Urrutia, K., Betancourt, D., Villamizar, M., Enabling Numerical Simulation and Real-Time Production Data to Monitor Water-Flooding Indicators, SPE 163811, 2013, Presented at the *SPE Digital Energy Conference*, The Woodland, TX, 2013. https://doi.org/10.2118/163811-MS
16. Yunus, K., Chetri, H., Saputelli, L., Waterflooding Optimization and Its Impact Using Intelligent Digital Oilfield (iDOF) Smart Workflow Processes: A Pilot Study in Sabriyah Mauddud, North Kuwait, IPTC-17315-MS, 2004, Presented at *International Petroleum Technology Conference*, Doha, Qatar, 2014. https://doi.org/10.2523/IPTC-17315-MS
17. Feder, J., Prescriptive Analytics Aids Completion Optimization in Unconventionals, *Journal of Petroleum Technology*, 2020, Vol. 72(4), pp 52–53. https://doi.org/10.2118/0420-0052-JPT
18. Flores, S., Linhares, S., Coletta, C.J., Landinez, G., Rios, R., Medina, Y., Luquetta, H., Drilling Performance Initiative in Campos Basin Block C-M-592, OTC-22511-MS, 2011, Presented at *Offshore Technological Conference*, Brazil, 2011. https://doi.org/10.4043/22511-MS
19. Moises, G.V., Rolin, T.A., Formigli, J.M., GeDIg: Petrobrass Carporate Program for Digital Integrated Field Management, SPE-112153, 2008, Presented at *SPE Intelligent Energy Conference and Exhibition held in Amsterdam*, Amsterdam, The Netherlands, 2008. https://doi.org/10.2118/112153-MS
20. Van den Berg, F., Perrons, R.K., Moore, I., Schut, G., Business Value from Intelligent Fields. Society of Petroleum Engineers, SPE-128245-MS, 2010, Presented at *SPE Intelligent Energy Conference and Exhibition*, Utrecht, The Netherlands, 2010. https://doi.org/10.2118/128245-MS
21. Almadi, S.M., AL-Khabbaz, F.M., and Abualsaud, Z.A., Digital Oil Field Cyber Security Best in Class, SPE-176754-MS, 2015, Presented at the *SPE Middle East Intelligent Oil & Gas Conference & Exhibition held in Abu Dhabi*, Abu Dhabi, 2015. https://doi.org/10.2118/176754-MS
22. Bazzana, A.M., Driving Down Costs in a Digital Oil and Gas Future. https://www.toptal.com/finance/energy-sector-expert/digital-oil-and-gas
23. World Economic Forum (2017), Digital Transformation Initiative: Oil and Gas Industry [White Paper]. http://reports.weforum.org/digital-transformation/wp-content/blogs.dir/94/mp/files/pages/files/white-paper-2017-dti-oil-gas.pdf
24. Marketwatch (2020, 5 November), Press Release. Retrieved 14 August 2021, from https://www.marketwatch.com/press-release/global-digital-oilfield-market-size-share-value-and-competitive-landscape-2020-2021-05-11?tesla=y
25. Globe Newswire (2021, 8 June), Intrado – Globe Newswire. Retrieved 19 August 2021, from https://www.globenewswire.com/en/news-release/2021/06/08/2243893/0/en/Digital-Oilfield-Market-to-Garner-USD-28-61-Billion-by-2027-Market-Research-Future-MRFR.html

26. Amin, S.O., Siddiqui, M.S., Hong, C.S., Building Scalable and Robust Architecture for Ubiquitous Sensor Networks with the Help of Design Patterns, *International Conference on Multimedia and Ubiquitous Engineering (MUE'07)*, 2007, pp 1046–1050. https://doi.org/10.1109/MUE.2007.101
27. Biz4intellia (2020), *Biz4Intellia.* Retrieved 20 August 2021, from https://www.biz4intellia.com/blog/pipeline-leak-detection-with-iot-in-oil-and-gas/
28. Latif, G., Alghazo, J.M., Maheswar, R., Sampathkumar, A., Sountharrajan, S., *IOT in the Field of the Future Digital Oil Fields and Smart Wells: Internet of Things in Smart Technologies for Sustainable Urban Development, Book Series (EAISICC)*, 2020, pp 1–17. http://doi.org/10.1007/978-3-030-34328-6_1
29. Atanas, J.P., Rodrigues, C.C., Simmons, R.J., Lean Six Sigma Applications in Oil and Gas Industry: Case Studies, 2016, *International Journal of Scientific and Research Publications*, Vol. 6(5), pp 540–544.
30. Rahmanifard, H., Plaksina, T., Application of Artificial Intelligence Techniques in the Petroleum Industry: A Review, 2019, *Artificial Intelligence Review*, Vol. 52, pp 2295–2318. https://doi.org/10.1007/s10462-018-9612-8
31. Wu, W., *Oil and Gas Pipeline Risk Assessment Model by Fuzzy Inference Systems and Artificial Neural Network*, 2015, The University of Regina, Faculty of Graduate Studies and Research. https://www.semanticscholar.org/paper/OIL-AND-GAS-PIPELINE-RISK-ASSESSMENT-MODEL-BY-FUZZYWu/91a0e3013292946860ec4fb7c2f7f19c9cf95524
32. Ailworth, E., Fracking 2.0: Shale Drillers Pioneer New Ways to Profit in Era of Cheap Oil, 2017, *The Wall Street Journal.* https://www.pressreleasepoint.com/fracking-20-shale-drillers-pioneer-new-ways-profit-era-cheap-oil

3 Big Data Analytics in Oil and Gas Industry

Vrutang Shah
Pandit Deendayal Energy University Gandhinagar, Gandhinagar, India

Jaimin Shah
The Maharaja Sayajirao University of Baroda, Vadodara, India

Kaushalkumar Dudhat
Pandit Deendayal Energy University Gandhinagar, Gandhinagar, India

Payal Mehta
Nirma University, Ahmedabad, India

Manan Shah
Pandit Deendayal Energy University, Gandhinagar, India

CONTENTS

DOI: 10.1201/b23013-3

3.1 INTRODUCTION

An oilfield is a commercial asset similar to any other in that it requires investment to generate revenue. When it comes to the way the oil and gas (O&G) sector spends to create cash flow, there are many specific factors to consider. These include operational hazards, a lengthy timeframe, and substantial responsibilities both during and after production ceases. However, the main investment goal is to maximise the field's economic return. To do this, technical and economic disciplines must be examined for areas where big data analytics may have a substantial and beneficial effect on performance. Additionally, the petroleum industry has entered an age of digitisation and intelligence as a result of the fast growth of oilfield industrial exploration and development technology and the continual improvement of automation and informatisation. Therefore, its data volume has quickly increased from the MB level to the TB or even PB level, demonstrating exponential development (Hassani & Silva, 2018; Hyne, 2019). Using China National Petroleum Corporation as an example, about 70 large-scale information systems have been developed and deployed after the "Tenth Five-Year Plan" and "Twelfth Five-Year Plan." The "exploration and production technologies data management system" alone handles about 1500 TB of data when it is operational (Hong-Qing et al., 2021).

During the exploration phase, it is critical to use a big data strategy to process the seismic data. Seismic data processing involves both high-performance computing and distributed and parallel high-performance data storage. This infrastructure is regarded required in order to build three-dimensional geological models that can be used to understand the complicated geological formations underneath the earth's surface (Vega-Gorgojo et al., 2016). Numerous studies have been conducted in recent years to gain a better understanding of subsurface petrophysical, reservoir, and fracture properties (Huang & Chen, 2019; Wang et al., 2012; Zhao et al., 2018; Zhou et al., 2017; Zhou et al., 2019), and to better characterise petrophysical properties, big data analytics has been used (Roy et al., 2013; Zhang et al., 2018; Zhao et al., 2014; Zhou et al., 2019). Additionally, numerous research and review articles demonstrate the potential benefits of using big data tools to seismic analysis in the petroleum sector (Dai et al., 2019; Luo et al., 2018; Noshi et al., 2018; Udegbe et al., 2018; Yadranjiaghdam et al., 2017). Seismic data centres gather and store data totalling 20 petabytes, which is almost 1000 times the amount of the United States Library of Congress's data (Beckwith, 2011; Hems & Perrons, 2013). Thus, technically feasible big data analytical tools can assist in reducing drilling lag time, many technical drilling and rig-related risks, and increasing drilling success rates (Technavio, 2015).

Drilling is the most critical component of the O&G sector, as it is the most difficult, dangerous, and expensive process. This accounts for almost half of the overall well expenditure; the drilling process requires 40% of time, while the other 60% of time is spent resolving drilling problems, process defects, rig issues, and latency periods (Basbar et al., 2016; Ngosi & Omwenga, 2015). According to data, this cost reaches about 100 million, and modern horizontal well drilling takes three weeks to complete drilling phase if complications do not occur (Marr, 2015). To minimise environmental hazards due to drilling operation, big data plays a vital role. For example, many sensors collect real-time rig data, which is feed and analysed via big data analytical tools. This will help to identify the drilling anomalies, enabling quick decision-making to halt operations and avert substantial environmental concerns (van Rijmenam, 2013). In the O&G industry, big data plays an important role in assisting the exploration and production for executing the successful drilling operation. Given the ever-changing nature of the sub-surface environment, big data is necessary for analysing and improving the critical operation, data collecting from geology, and operation aid in improving the drilling operation.

Modern Enhanced Oil Recovery (EOR) can possible due to the big data, which collect data from sensors and analyse through big data analytics that enable the productivity of mature wells that were previously deemed to have reached their maximum production to be increased. Indeed, a 1% increase in output from mature fields that are currently operational may add two years to the world's supply of oil and gas (Cowles, 2015). To increase output and recovery, digital interventions like as artificial lift technologies, reservoir monitoring and other oil recovery techniques are increasingly being deployed. Often, a combination of information and first principle-based processes is beneficial for increasing oil recovery from the industry norm 20% to a more aggressive 50% (MoPNG, 2020). Optimising oil recovery from existing wells is a critical priority for the oil and gas industry. By combining analytics with a variety of big data sources—seismic, drilling, and production data—reservoir engineers can trace reservoir evolution through time and assist production engineers with decision-making when changing lifting techniques. This method may be advantageous for hydraulic fracturing in shale gas areas (Baaziz & Quoniam, 2014).

Natural gas and crude oil production are often transported via long-distance metallic pipes. Corrosion is a significant source of pipeline problems due to the nature of the environment and high temperatures (Mohamed et al., 2015). Almost 30% of pipeline problems are caused by exterior corrosion (Nicholson, 2007). These pipeline flaws can result in significant financial losses, environmental degradation and loss of life. According to the most recent data available in 2015, the world's pipeline network totals more than 3.5 million kilometres. Numerous additional pipelines totalling hundreds of kilometres are being planned or constructed (Mohamed et al., 2017). Currently, the majority of oil and gas storage and transportation companies monitor the safety of their storage systems and pipelines using a three-tiered approach that collects temperature, pressure, and other data from long-distance pipes and transmits it in real time to a data monitoring centre via the Internet, where it is analysed and processed in real time for possible early warning situations (Kong et al., 2020).

Numerous smart sensors are utilised in the contemporary O&G sector, pipelines, and processing plants, resulting in the world's most dense and biggest Internet of Things (IoT) networks. This Internet of Things is the backbone of the upstream oil and gas business, as well as the pipeline industry, which is responsible for refining and marketing processed goods (Anderson, 2017). Light crude oil, naphtha, kerosene and heavier crude oil are refined from crude oil. When demand for lighter goods surged, they turned to a newly developed technology called thermal cracking to transform the heavier crude oil into a lighter one. Refineries may use big data to optimise their energy and fossil fuel use, as well as numerous other elements of their operations, such as maintenance and repair. Big data helps predictive maintenance in the LNG and CNG industries, minimising unplanned shutdowns and operations, fixing issues, and strengthening infrastructure (Patel et al., 2020).

By examining the growing volume of data captured by the oil and gas industry, they can overcome industry-specific challenges such as making complex operational processes visible, enabling performance optimisation, simplifying equipment life cycle management, simplifying logistical challenges, and making compliance with environmental regulations convenient. Big data analytics may be an excellent tool for enhancing individual and corporate productivity, economic circumstances, and safety. This chapter examines the application of big data analytics in petroleum engineering, citing several academic articles and company case studies from the field implementation process. The chapter begins with an overview of big data analytics. The second section discusses how big data analytics is being used in various divisions of the petroleum upstream and downstream industries. Additionally, the concluding portion discusses the next steps required for integrating big data analytics and the obstacles that must be addressed.

3.2 OVERVIEW OF BIG DATA ANALYTICS

Big data (BD) is a burgeoning sector with applications in a wide variety of disciplines. It is critical since it ensures optimal operation efficiency, price utilisation, security, and consumer perspectives. As the term "big data" implies, it belongs to a vast amount of structured, semi-structured and unstructured datasets. Structured data are data which have labels, whereas unstructured data are unlabelled data. The O&G industry's data are mostly unstructured, which makes analysis extremely difficult. The O&G industry produces huge volumes of data daily as a result of current technologies. Textual drilling reports and well logs, and also AutoCAD drawings, are examples of unstructured data sources in the O&G business (Mohammadpoor & Torabi, 2020). Sensors installed in the drilling process gather data at each time step, which may be as little as 5 seconds, allowing for the collection of vast amounts of data. Unstructured data analysis technology is still in its infancy, making it difficult to analyse. Garter (jpt.spe.org/oil-and-gas-has-problem-unstructured-data) asserts that the amount of data generated will rise by 800% over the next half-decade, despite the fact that 80% of data are unstructured.

As mentioned above, data collecting is critical in the O&G business, and given the rapid rate at which data are gathered by various types of sensors, which can take any form, the next step is data storage, management, and analysis. This will have a significant financial impact on the O&G sector. This is required, however, if the plant's efficiency is to be increased or if geographical data analysis is to be performed. Previously, they required servers to hold this volume of data, but, as a result of the technological change, data can now be stored simply via Cloud Computing, IoT (Internet of Things) and Fog Computing (Mounir et al., 2018). The growth of digital oilfields, in which multiple sensors and recording devices create millions of data points daily in every field of drilling and geographical location, upstream and downstream, has expedited the rise of BD in the O&G sector. Several of the most challenging issues in digital oilfields are with data transportation from the field to data processing centres, which is highly reliant on the type of data, its volume, and its protocol (Neri, 2018). Additional issues such as data privacy and information confidentiality develop as a result of this volume of data, necessitating consideration of cyber-security considerations. Cyber attackers have increased their regularity and sophistication in recent years, targeting a variety of organisations and businesses. As a result of the inexorable exponential rise of cyber–physical systems, security problems are exacerbated (Nguyen et al., 2020).

Data gathering is essential for a variety of upstream purposes, including exploration and reconnaissance, drilling, reservoir engineering, and production engineering. It is used in the midstream for pipeline and ship transport, and in the downstream for refining, health and safety, trade, and sales. Scouting and exploration data sets often contain seismic data, microseismical data, and 1D, 2D, and 3D geological maps. Drilling statistics contain information on the efficiency and performance of drilling rigs. Data are used in reservoir engineering to manage reservoirs, to close the loop on reservoir management, to integrate asset modelling, to sequester CO_2 and to improve oil recovery. Data from Production Engineering is used for a variety of purposes, including automated decline analysis, production allocation methodologies, electric submersible pumping, rod pump wells, hydraulic fracturing projects, field development, forecasting hazardous event forecasting, and well casing damage prediction. Ship transport data are used to analyse the performance and energy efficiency of ships, whereas pipeline transport data are used to monitor pipelines. Additionally, data are gathered for the management O&G assets, comprehensive refinery operations, and increased refining sector productivity. The Health and Safety Executive uses the data it collects to improve workplace safety, predictive safety analytics, and disease mitigation. Finally, trade and sales data are analysed in order to forecast crude oil prices and determine market volatility (Nguyen et al., 2020).

The petroleum sector encompasses the exploration, drilling, production, transportation, refining, and sale of hydrocarbons, all of which generate a huge amount of data. Additionally, safety and health are critical components of this industry's operations; without them, the company risks posing major ergonomic, economic, and environmental hazards. Additionally, log data obtained during drilling are difficult to manage and exhibit a significant degree of variability. The petroleum sector can regulate these large amounts of data, shown in Table 3.1, through the use of big data technology, but this can be costly.

The 6Vs that make this BD tool feasible are volume, variety, velocity, veracity, value, and variability (Alguliyev et al., 2017). Petabytes to gigabytes of data are included in O&G. The term "variety" alludes to the fact that data may be found in a number of formats, including structured, unstructured, and semi-structured. Velocity is a term that relates to the rate at which data is created throughout each time step. Veracity refers to the data analysis process. When the data quality exceeds this, the process becomes more efficient. Variability is a term that refers to the changes that occur in data throughout processing and during its existence. All Vs data adds value in this way by exposing the forecasting of possible geological difficulties and detecting failures before to their occurrence. This has significant implications for production, failure detection, efficiency enhancement, health optimisation, and the future trend of O&G in the stock market, among other things. By combining this data with advances in massively parallel computer machines, increased storage capacity, and a

TABLE 3.1
The Amount of Data Captured by Various Sectors of the O&G Business (Lu et al., 2019)

Data Collection Section	Amount of Data
Drilling	0.3 GB/well (daily)
Submersible pump monitoring	0.4 GB/well (daily)
Wireline	5 GB/well (daily)
Seismic	100 GB – 2 TB/survey
Plant process	4-6 GB (daily)
Pipeline inspection	1.5 TB (600 km)
Plant operational	8 GB (annually)
Vibration	7.5 GB (per customer annually)

new generation of wireless networks, real-time applications such as remote oilfield monitoring have become more practical. Advanced BD analytics technologies can help businesses optimise the output potential of their assets while also addressing performance shortcomings (Brun et al., 2017).

Recent studies have reported the value of BD implementation throughout the O&G industry. This section highlights many noteworthy uses of BD in the oil and gas supply chain. Recent technological developments in seismic devices have significantly expanded the amount of data accessible in upstream oil and gas. BD analytics has transformed into an economical method for managing and analysing this data. For instance, some researchers have substituted the BD approach and Hadoop platform for conventional tools in order to analyse huge seismic datasets, find key geological features, classify the reservoir, and define geological concerns (Joshi et al., 2018).

3.3 BIG DATA ANALYTICS IN UPSTREAM SECTOR

3.3.1 Petroleum Exploratory

Petroleum Exploration is the first step towards producing the economically effective petroleum production from the purchased/opting to bid area from government regulation. This pivotal stage mandates the necessary action required, on which multi-million-dollar investment is based, to predict the basin's potential for the economic feasibility of petroleum production. Here, big data analytics can help us extract most useful information from the data on which to base the best bid decision and plan for effective exploitation of the area.

Big data analytics can aid in the following areas of Petroleum Exploration:

1. Assess the prospect's identity: Geospatial data analysis, various oil and gas reports, and syndicated feed all aid the company in submitting a competitive bid for the prospect (Hems et al., 2013).
2. Extracting valuable data: Seismic trace identification enables the most value to be extracted from the enormous amount of available seismic

data by identifying potential traces that effectively understand the subsurface and plan an accurate drilling programme (SEAWANDERfER, 2020; Zhao, 2018).
3. Enhance the subsurface model in advance of short- and long-term reservoir operations: Actual drilling and logging data enables the modification of existing seismic data and the construction of an accurate subsurface model for use in future plans based on current conditions. As a result, there is a greater likelihood of increased revenue generation (Kozman & Holsgrove, 2019).

The finding of new hydrocarbon resources requires substantial capital, people, and logistical expenditures. Furthermore, given the high expense of digging a deepwater oil well often exceeding $100 million, looking in wrong location will be unhelpful and tedious. Shell circumvents this problem by sending data through fibre optic connections to its servers hosted by Amazon Web Services (AWS). This provides engineers with a far more precise picture of what lies underneath and saves significant time and effort (Zaidi, 2017). When combined with increased integration of diverse sources of seismic and reservoir data, these advancements have the potential to significantly improve decision-making for exploratory drilling, reservoir modelling, and development.

3.3.2 Drilling and Completion Optimisation

Drilling is a necessary, hazardous, tough, and expensive activity in the O&G business. While drilling accounts for more than half of all well expenditures, it accounts for just 42% of time. The remaining 58% is spent on drilling challenges, rig movement, defects, and latency delays. During the Drilling and Completion Optimisation phase, the primary goal of big data analytics is to reduce drilling time and consequently costs, while increasing safety. Large and small operators alike have built real-time operations support centres through the use of current technologies and the skills and expertise of a widely distributed well engineering organisation (Lu et al., 2019).

Big Data analytics can aid in the following areas of Drilling Engineering:

1. Develop and continuously improve the drilling model: Using pre-existing well data and real-time drilling data, big data analytics predicts the most accurate drilling path, maximising the benefit of the drilling operation with the least amount of damage. As a result of the real-time data, the drilling model is updated, assisting in the optimisation of the drill parameters. By identifying the factors that contribute to NPT (non-productive time), big data analytics assists in reducing NPT (Evensen & Haaland, 2019; Flores et al., 2011).
2. Safety enhancements and drilling optimisation: Real-time drilling data enables big data analytics models to anticipate and avoid undesirable situations such as stuck pipe, kicks and, in the worst-case scenario, blowouts (JPT, 2020).

3. Predictive Maintenance: forecast drill downtime/maintenance (Baaziz & Quoniam, 2014; NS Energy, 2016).
4. Reduce well-planning preparation time while increasing reliability in order to maximise well productivity at a reduced cost: Using real-time drilling data, the digital twin enables real-time modification of various parameters used in the drilling process (Baker, 2020).

To maximise oil and gas production while minimising environmental impacts, O&G businesses could use big data. The gathering and analysis of real-time data allows early detection of drilling abnormalities, allowing for timely shut-downs to avoid significant environmental hazards (van Rijmenam, 2013). Petro-Hunt leverages data digitisation to streamline its procedures and policies, as well as its automated and simultaneous well engineering workflows and drilling programme. These steps create the situation which improves the well-planning programme by mitigating the problems associated with the well-planning process in HPHT wells. The outcome of the project shows 50% reduction in well-planning time, improving cross-discipline collaboration and safety by anti-collision scanning (Schlumberger, 2018). This great step towards the digitalisation makes the psychological adaptation of the big data analytics in the O&G industry and makes it possible for other companies.

3.3.3 Reservoir Characterisation, Simulation, and Engineering

Reservoir characterisation is the method of constructing a model of a subsurface body of rock that combines all of the reservoir's unique characteristics associated with its hydrocarbon accumulation capacity. It is critical for conventional reservoir management because it enables upstream engineers to make educated decisions about the extraction of oil and gas from these assets. The reservoir characterisation process entails assessing reservoir parameters, which are then used in computer modelling (for reservoir simulation) to estimate the behaviour of the fluids flowing through the reservoir under various production scenarios. The ultimate objective is to develop a portfolio of the most efficient oil production options. The effectiveness of drilling, completion, and production techniques is dependent on reproducible reservoir characterisation accuracy throughout the exploration and production (E&P) value chain.

Big data analytics can aid in the following areas of Reservoir Engineering:

1. Enhance engineering studies and gain a better understanding of the subsurface: Big data analytics enables the creation of sophisticated subsurface models, which enables the identification of commercial prospects earlier (Feblowitz, 2012; MoPNG, 2020). Additionally, the number of active man-hours required to create the reservoir model is reduced, which increases labour productivity.
2. Effective reservoir and asset management and optimisation: Integrating drilling data into the reservoir model enables a more complete understanding of the reservoir. This knowledge enables reservoir engineers to make

more informed decisions about how to optimise the number of wellheads and extract the maximum value from the reservoir (Baaziz & Quoniam, 2014; MoPNG, 2020).

3. Improve hydraulic fracturing job performance: Various Neural Networks and Support Vector Machine algorithms can be used to optimise fracturing parameters and create the best data-driven hydraulic fracturing model possible (Temizel et al., 2015).
4. Increase the efficiency of EOR projects: Big data analytics enables the identification of the best candidate for EOR techniques, thereby reducing reservoir management uncertainty and optimising hydrocarbon recovery potential. Additionally, the EOR process design can be optimised to reduce operating costs and increase project output (Mohaghegh et al., 2014).

A hybrid model combining machine learning and hard computing techniques speeds up the linked wellbore hydraulic and numerical reservoir simulation procedures. According to a research, when the ANN-numerical model is employed for gas lift injection, it exhibits a 5% divergence from the complete conventional numerical model while reducing computation time by 160 (Putcha & Ertekin, 2018). Although the divergence may vary for different models, the decrease in computing time shows the enormous accomplishment of integrating big data analytics into reservoir modelling.

3.3.4 Production Engineering

The primary objective of a production engineer is to maximise petroleum production, which is dependent on subsurface physics and chemistry. Thus, data collection from exploration and drilling activities is critical in determining the optimum rate. Additionally, petroleum production is dominated by mature/declining fields, accounting for 70% of total production (Hassani & Silva, 2018).

Big data analytics can aid in the following areas of Production Engineering:

1. Increase oil recovery: Using big data analytics on seismic, drilling and available production data, reservoir engineers can develop a new optimal oil recovery strategy, and production engineers can choose the optimal lifting methods for the particular case using the supporting data (Baaziz & Quoniam, 2014; Feblowitz, 2012).
2. Increased overall field recovery through the use of production forecasting: Using big data analytics, thousands of wells from an asset can be analysed, assisting in identifying wells with production below the threshold limit. This instructs the production engineer to immediately take corrective action to restore production (Feblowitz, 2012).
3. Increased safety and risk mitigation: Early detection of problems through the use of big data analytical methods enables production engineers to identify issues earlier, before serious problems such as gas breakthrough and slugging occur (Baaziz & Quoniam, 2014).

4. Early prediction of ESP failure: By identifying the complex relationships between various parameters of the ESP working system in comparison to previous failure events, early prediction of ESP failure is possible approximately two weeks prior to the actual failure event. Additionally, this minimises equipment wear and tear (Bhardwaj et al., 2019).
5. Overall Work Productivity: Using Data analytics techniques and Bayesian Belief Networks as reasoning tools, an automated workflow candidate selection process aids in decision-making for selecting the ideal candidate for workover. Additionally, this method identifies non-value-added or low-value-added work through automation, enhancing overall economic benefits through effective selection of workover candidates and enabling employees to derive significant monetary benefits from their work. Thus, unplanned downtime, the primary source of revenue loss in the O&G industry, can be avoided by instructing engineers to select the appropriate maintenance programme schedule (MoPNG, 2020).
6. Additional benefits include cost transparency, efficient resource allocation and reporting, increased overall work productivity, and supply chain management (MoPNG, 2020).

Petroleum Development Oman creates the Machine Learning algorithm to predict the relation between various design and operating variables in the Electric Submersible Pump using the input data. This digital smart system monitors, control, and optimise the oil production behaviour. Additionally, reliable prediction of the failure of the system is possible using the algorithms before the actual failure takes place (Awaid et al., 2014). This smart system gives the confidence and real-life proof of the advantage to provide the most optimise recovery and maintenance schedule possible, before it impacts the company workflow.

3.4 BIG DATA ANALYTICS IN MIDSTREAM AND DOWNSTREAM SECTOR

3.4.1 Oil and Gas Transportation

In oil and gas industry, the transportation of the oil and gas is primarily a major concern due to its complex nature with minimum risk. Oil tankers, barges, pipelines, rail transport and trucks have been the major transportation mode in oil and gas industry recently. Crude oil and its product, natural gas, and oversized equipment all require special handling with proper safety regulations. In oil and gas supply chain, cargo security, storage and transportation equipment and delivery speed play a notable role. The main challenge behind the oil and gas transportation is to choose the mode of transportation for successful delivery. There are still many problems associated with the oil and gas storage and transportation processes with the increasing demand of the oil and gas resources. Still, the key issues that exist in different modes of transportation as well as in oil and gas storage equipment require special attention to solve in an efficient way (Hua et al., 2015).

O&G corporations use complex algorithms to analyse economic considerations, storage and transportation costs, and varied weather patterns in order to determine how

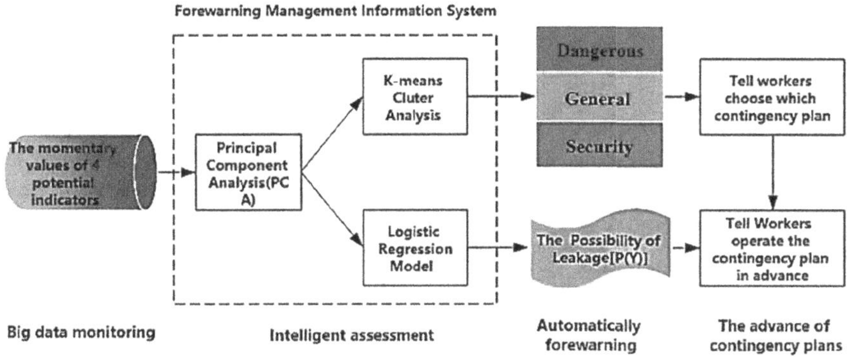

FIGURE 3.1 The management mode of FMIS (Wang et al., 2017).

and where to deliver refined oil and gas products (Bekker, 2020). The industry makes extensive use of sensor analytics to ensure the safe transportation of oil and gas. The research continues to improve transportation through the use of big data analytics from previous years (Mohammadpoor & Torabi, 2020). To improve shipping performance, big data analytics is applied. Propulsion power is used to enhance shipping performance via big data applications. The primary advantage of propulsion power is that it reduces greenhouse gas emissions. Data analytics is accomplished through the use of neural network models such as eXtreme Gradient Boosting (XGBoost) and multilayer perception (MLP). The sensor data collected throughout an LCTC (Large Car Truck Carrier) M/V (Anagnostopoulos, 2018). Various predictive maintenance software applications are used to analyse sensor data from pipelines and tankers in order to detect abnormalities and thereby avoid accidents (Brancaccio, 2016).

A method for evaluating pipeline efficiency is developed by monitoring changes in energy and transmission volume using the energy balance principle. The operating efficiency of the pipeline and associated facilities is determined using a Data Envelopment Analysis (DEA) model. The DEA model employs an analytic hierarchy process (AHP) to quantify the effect of various facilities (Fan et al., 2019). Numerous studies on the hysteretic character of contingency plans and the risk assessment of oil and gas transportation and storage facilities have been conducted. Using K-means cluster analysis and logical regression fitting is one of the ways. Because the logistic regression model is applicable to a wide variety of applications, it may be used to scientifically analyse the risk level associated with a particular piece of storage and transportation equipment (T. Wang et al., 2017). Figure 3.1 illustrate that the Forewarning Management Information System (FMIS) is used to predict the risk of storage facilities and give information in advance to operate contingency plan to avoid the hazards.

3.4.2 Refining

The refining sector accounts for 4% of total worldwide primary energy consumption, making it critical to improve refinery energy efficiency. In the early 1970s, transformation of the board-mounted and pneumatic refinery controls to the electronic distributed control system (DCS) had become popular which helped refinery operations to improve the production, optimisation of the product and reliability. The information architecture in today's modern refineries has thrived and has been stable at some extent from the last several years. However, much of the data remained locked away in proprietary systems still today and referred as a "dark data". Dark data often include the data of the equipment monitoring, specialised chemical programmes, inspection readings, maintenance data, and non-core refinery operations. From the last several years, "Industrial Internet of Things" has been under development which evolved new sensors with high spans, controllers and different data handling infrastructures. With this development, it allows the more rapid, transparent remote measurements, analytic algorithm in automated application that leads to make impossible things to do possible. The improvement in performance evolution and analysis enhances the opportunities to optimise profitability and reliability.

The laboratory information system and the asset management system are the primary data sources in the refining business. O&G firms leverage data management solutions and big data predictive analytics to increase operational efficiency, decrease equipment downtime and maintenance costs, and ultimately improve petrochemical asset management. On historical data, a three-stage approach is used to analyse performance and optimise petrochemical asset management. For instance, in the case of a four-stage fractured gas compressor (CGC), a three-stage big data analytics approach was used. The first stage involved forecasting CGC's performance using historical and current operating data. The second stage involved fine-tuning the performance prediction algorithm in accordance with the device's failure state end-of-life criteria. Finally, a visually appealing and user-friendly report summarising the estimated performance was delivered to management for decision-making purposes. These analytic tools established through the use of big data applications have the potential to drastically lower maintenance and downtime expenses (Moritz von Plate, 2016). Big data analytics is being implemented in one of Spain's integrated refineries. Google Cloud would deliver consultation and data analytics products with (ML) machine learning services to Repsol as part of this collaboration (Brelsford, 2018).

Big data analytics is used to optimise pricing and financial risk. The refining business maximises revenues by tracking and analysing large amounts of data. Big data management software integrates the Hadoop file system for Big data analysis. Big data analysis requires a solid storage infrastructure that allows for constant access to the data. The downstream sector—processing, logistics, and sales—is where Big data analytics is expected to make a difference. It contributes to estimating demand for oil goods in retail sales markets and to analysing regional rivals' pricing fluctuations (Brancaccio, 2016).

The two models utilised for refinery energy analysis are DEA and Principal Component Analysis (PCA). The DEA approach is used to ascertain the efficiency of a particular dataset of decision-making units (DMUs). There are two sorts of DEA models: input-oriented and output-oriented. Input-oriented DEA models are used in the refinery application because the efficiency of refinery operations is evaluated using a fixed structure. The input-oriented DEA model enables the evaluation of refineries' performance in terms of producing refinery products with the least amount of power and fuel. The DEA's numerous models include the CCR model, the Andersen model, and the Petersen model. Generally, the Andersen and Petersen model is employed in the refining business. CCR is not widely utilised in the refining business. Because it uses the same index for all decision-making units, it is unable to give rank-efficient units to all decision-making units. If the value of the decision-making unit's efficiency score is equal to or greater than one, the Andersen and Petersen model is significantly more efficient. PCA is mostly used in multivariate statistics to reduce the number of variables in all decision-making units. PCA is utilised in a variety of settings, including medical facilities, companies, and communities. PCA generates the outputs by combining the inputs from numerous sources. PCA analysis is used to determine the concentrations of air contaminants in the Athabasca oil sands (Patel et al., 2020).

Big data analytics is also employed in prognostic foresight. Prognostic analytics techniques assist in more efficiently analysing data for forecasting than predictive analytics techniques. Prognostic analytics methods contribute to the refinery's improvement in a variety of areas, including operations, asset management, risk management, maintenance, and finance. Prognostic analytics provide long- and short-term maintenance scheduling, manpower planning, and allocation in a maintenance region. Prognostic analytics enables production schedules to be planned based on anticipated availability profiles in a particular sector of operations. Prognostic analytics creates a portfolio of strategies for lowering maintenance costs, increasing insurance policies, and increasing profit in the finance sector (M V Plate & Ag, 2018).

Big data analytics has the potential to significantly increase operational efficiencies, reliability, and productivity across the downstream sector. Big data analytics would enable incremental improvements in operational excellence. According to the refining sector's future plans, it is crystal obvious that Big data analytics would be applied to global refineries to improve their reliability and operations.

3.5 CHALLENGES, FUTURE STEPS, AND BENEFITS OF BIG DATA ANALYTICS IN OIL AND GAS INDUSTRY

Various challenges are prevailing in today's market. And for executing big data analytics in O&G industry we require to overcome those obstacles. Few of the challenges to create the infrastructure for big data analytics for proper implementation of the technology are outlined below.

1. Computer system cost (managing, storing and processing data).
2. Digitalisation of the oil field requires the use of many sensors and generates a large quantity of data, which presents a problem in terms of data transmission and storage according to the kind, volume, and protocol of data (Gidh et al., 2016; Mohammadpoor & Torabi, 2020; Neri, 2018).
3. A lack of awareness and support for business, as well as staff knowledge and competence in the diverse fields of big data and petroleum (Feblowitz, 2013).
4. Constraints associated with recording sensors (frequency and quality) (Maidla et al., 2018).

5. Processing and analysis of geological and geophysical data need the use of a variety of software programmes (single platform is not yet available) (Aliguliyev et al., 2016).
6. Each data cluster has a significant volume of problem-oriented data (which requires more pre-processing) (Aliguliyev et al., 2016).
7. A silo mindset prevails among members of a multidisciplinary team (Pandey et al., 2020).
8. Psychological variables affecting the shift to newer technologies and the integration of change into the regular course of business (Pandey et al., 2020).

Future Steps Required for the proper implementation of big data Analytic in O&G Industry are outlined below (Srinivasan, 2018).

1. It is necessary to standardise data storage technologies.
2. IoT device planning for data transfer to a cloud storage device.
3. Customer segregation from the big data analytics service provider is needed by establishing a cloud account for the customer.
4. A platform for multi-client collaboration is needed, since petroleum E&P operations are often not carried out by a single licence holding firm.
5. Educating big data system designers on the fundamentals of big data implementation (petroleum technical expertise).

Overall benefits and future scope of implementation of the big data analytics in O&G Industry are outlined below (Holdaway, 2014).

1. Shortening the time required to discover the first sign of oil.
2. Throughout the asset's lifetime, its productivity improves.
3. Embedding sophisticated business intelligence and analytics.
4. Ensuring that the workforce has access to the appropriate information at the appropriate time.
5. Improving the outcomes of planning and forecasting.
6. Increase in the business cycle's transparency.
7. Increase in the total income produced by the specific asset/field in comparison to the previous year.

3.6 CONCLUSION

Many oil and gas companies started their project regarding the implementation of big data analytics which shows promising results to further extend the field to all the operations of this industry. The digitalisation process towards the effective mechanism and management of the project, using big data analytics, shows tremendous potential and achievable goal for the future operations. These steps lead to the creation of the new job opportunity, along with reorganisation of the organisation structure for very effective and safe business operation for sustainable competitive advantage. There is main gap towards the adaptation of big data analytics which is mentioned in the chapter earlier. Filling this gap is essential for the complete effective transformation of the oil and gas companies.

REFERENCES

Alguliyev, R. M., Aliguliyev, R. M., & Hajirahimova, M. S. (2017). Big Data Integration Architectural Concepts for Oil and Gas Industry. *Application of Information and Communication Technologies, AICT 2016 – Conference Proceedings*. https://doi.org/10.1109/ICAICT.2016.7991832

Aliguliyev, R. M., Imamverdiyev, Y. N., & Abdullayeva, F. J. (2016). The investigation of opportunities of big data analytics as analytics-as-a-service in cloud computing for oil and gas industry. *Problems of Information Technology*, 9–22. https://doi.org/10.25045/jpit.v07.i1.02

Anagnostopoulos, A. (2018, June 10). Big Data Techniques for Ship Performance Study. onepetro, *The 28th International Ocean and Polar Engineering Conference*, Sapporo, Japan.

Anderson, R. N. (2017). Petroleum Analytics Learning Machine' for Optimizing the Internet of Things of Today's Digital Oil Field-to-Refinery Petroleum System. *Proceedings – 2017 IEEE International Conference on Big Data, Big Data 2017, 2018-January*, 4542–4545. https://doi.org/10.1109/BIGDATA.2017.8258496

Awaid, A., Al-Muqbali, H., Al-Bimani, A., Yazeedi, Z., Al-Sukaity, H., Al-Harthy, K., & Baillie, A. (2014). ESP Well Surveillance Using Pattern Recognition Analysis, Oil Wells, Petroleum Development Oman. *IPTC 2014: International Petroleum Technology Conference*. https://doi.org/10.2523/IPTC-17413-MS

Baaziz, A., & Quoniam, L. (2014). How to use big data technologies to optimize operations in upstream petroleum industry. *International Journal of Innovation*, *1*(1), 19–25. https://doi.org/10.5585/iji.v1i1.4

Baker, J. (2020, May 15). *Digital Twins are Delivering New Efficiencies for the Oil and Gas Industry*. NS Energy. https://www.nsenergybusiness.com/features/digital-twins-oil-gas/

Basbar, A. E. A., Al Kharusi, A., & Al Kindi, A. (2016). Reducing NPT of rigs operation through competency improvement: A lean manufacturing approach. *Society of Petroleum Engineers – SPE Bergen One Day Seminar*. https://doi.org/10.2118/180066-MS

Beckwith, R. (2011). Managing big data: Cloud computing and co-location centers. *Journal of Petroleum Technology*, *63*(10), 42–45. https://doi.org/10.2118/1011-0042-JPT

Bekker, A. (2020). *How to Benefit from Big Data Analytics in the Oil and Gas Industry*. https://www.scnsoft.com/blog/big-data-analytics-oil-gas

Bhardwaj, A. S., Saraf, R., Nair, G. G., & Vallabhaneni, S. (2019). Real-time monitoring and predictive failure identification for electrical submersible pumps. *Society of Petroleum Engineers – Abu Dhabi International Petroleum Exhibition and Conference 2019, ADIP 2019*. https://doi.org/10.2118/197911-MS

Brancaccio, E. (2016). *Big Data in Oil and Gas Industry*. Esds. https://www.esds.co.in/blog/big-value-big-data-oil-gas-industry/#sthash.eRZ9Zjt0.FaWDdTJz.dpbs

Brelsford, R. (2018). Repsol launches big data, AI project at tarragona refinery. *Oil & Gas Journal*. https://www.ogj.com/refining-processing/refining/operations/article/17296578/repsol-launches-big-data-ai-project-at-tarragona-refinery

Brun, A., Trench, M., & Vermaat, T. (2017). Why oil and gas companies must act on analytics. *McKinsey & Company*, 1–5. https://www.mckinsey.com/industries/oil-and-gas/our-insights/why-oil-and-gas-companies-must-act-on-analytics

Cowles, D. (2015, April 15). *Oil, Gas, and Data – O'Reilly*. O'REILLY. https://www.oreilly.com/content/oil-gas-data/

Dai, H. N., Wong, R. C. W., Wang, H., Zheng, Z., & Vasilakos, A. V. (2019). Big data analytics for large-scale wireless networks: Challenges and opportunities. *ACM Computing Surveys*, *52*(5). https://doi.org/10.1145/3337065

Evensen, O., & Haaland, Ø. (2019). *Artificial Intelligence Improves Real-Time Drilling Data Analysis*. Offshore. https://www.offshore-mag.com/drilling-completion/article/16764029/artificial-intelligence-improves-realtime-drilling-data-analysis

Fan, M. W., Ao, C. C., & Wang, X. R. (2019). Comprehensive method of natural gas pipeline efficiency evaluation based on energy and big data analysis. *Energy*, *188*. https://doi.org/10.1016/J.ENERGY.2019.116069

Feblowitz, J. (2012). The big deal about big data in upstream oil and gas. *IDC Energy Insights*, 1–11.

Feblowitz, J. (2013). Analytics in Oil and Gas: The Big Deal About Big Data. *Society of Petroleum Engineers – SPE Digital Energy Conference and Exhibition 2013*, 286–291. https://doi.org/10.2118/163717-MS

Flores, S., Linhares, S., Coletta, C. J., Landinez, G. A., Rios, R. O., Marquez, Y. M., De Oliveira, P. C., & Luqueta, H. (2011). Drilling performance initiative in campos basin block C-M-592. *Proceedings of the Annual Offshore Technology Conference*, 2, 796–803. https://doi.org/10.4043/22511-MS

Gidh, Y., Deeks, N., Grovik, L. O., Johnson, D., Arumugam, S., Schey, J., & Hollingsworth, J. (2016). WITSML v2.0: Paving the Way for Big Data Analytics Through Improved Data Assurance and Data Organization. *Society of Petroleum Engineers – SPE Intelligent Energy International Conference and Exhibition*. https://doi.org/10.2118/181096-MS

Hassani, H., & Silva, E. S. (2018). Big data: A big opportunity for the petroleum and petrochemical industry. *OPEC Energy Review*, *42*(1), 74–89. https://doi.org/10.1111/opec.12118

Hems, A., & Perrons, R. (2013, January 1). *Cloud Computing: What Upstream Oil and Gas Can Learn from Other Industries*. JPT. https://jpt.spe.org/cloud-computing-what-upstream-oil-and-gas-can-learn-other-industries

Hems, A., Soofi, A., & Perez, E. (2013). Drilling for new business value: How innovative oil and gas companies are using big data to outmaneuver the competition. *A Microsoft White Paper*, 1–13.

Holdaway, K. R. (2014). *Harness Oil and Gas Big Data with Analytics: Optimize Exploration and Production with Data-Driven Models*. John Wiley & Sons.

Hong-Qing, S., Shu-Yi, D., Yuan-Chun, Z., Yu-He, W., Jiu-Long, W., Hong-Qing, S., Shu-Yi, D., Yuan-Chun, Z., Yu-He, W., & Jiu-Long, W. (2021). Big data intelligent platform and application analysis for oil and gas resource development. *Chinese Journal of Engineering*, *43*(2), 179–192. https://doi.org/10.13374/J.ISSN2095-9389.2020.07.21.001

Hua, J., Li, H., & Liu, Y. (2015). Study on the problems and countermeasures of oil and gas storage and transportation systems in China. *Proceedings of the 2015 International Conference on Education Technology, Management and Humanities Science*, 27, 1269–1272. https://doi.org/10.2991/ETMHS-15.2015.279

Huang, C., & Chen, S. (2019). Stress analysis of an inclined borehole subjected to fluid discharge in saturated transversely isotropic rocks. *International Journal of Geomechanics*, *19*(11), 04019118. https://doi.org/10.1061/(asce)gm.1943-5622.0001503

Hyne, N. J. (2019). *Nontechnical Guide to Petroleum Geology, Exploration, Drilling & Production*. PennWell Books.

Joshi, P., Thapliyal, R., Chittambakkam, A. A., Ghosh, R., Bhowmick, S., & Khan, S. N. (2018). Big Data Analytics for Micro-Seismic Monitoring. *Offshore Technology Conference Asia 2018, OTCA 2018*, March, 20–23. https://doi.org/10.4043/28381-ms

JPT. (2020, February 25). *Halliburton Signs on to Support Pertamina's Digital Transformation*. https://jpt.spe.org/halliburton-signs-support-pertaminas-digital-transformation

Kong, W., Yu, J., Yang, J., & Tian, T. (2020). Model building and simulation for intelligent early warning of long-distance oil & gas storage and transportation pipelines based on the probabilistic neural network. *IOP Conference Series: Earth and Environmental Science*, 546, 22009. https://doi.org/10.1088/1755-1315/546/2/022009

Kozman, J. B., & Holsgrove, M. (2019). Maximising value from seismic using new data and information management technologies. *ASEG Extended Abstracts, 2019*(1), 1–5. https://doi.org/10.1080/22020586.2019.12073101

Lu, H., Guo, L., Azimi, M., & Huang, K. (2019). Oil and gas 4.0 era: A systematic review and outlook. *Computers in Industry, 111*, 68–90. https://doi.org/10.1016/J.COMPIND.2019.06.007

Luo, X., Bhakta, T., Jakobsen, M., & Nævdal, G. (2018). Efficient big data assimilation through sparse representation: A 3D benchmark case study in petroleum engineering. *PLoS ONE, 13*(7), e0198586. https://doi.org/10.1371/journal.pone.0198586

Maidla, E., Maidla, W., Rigg, J., Crumrine, M., & Wolf-Zoellner, P. (2018). Drilling Analysis Using Big Data Has Been Misused and Abused. *Society of Petroleum Engineers – IADC/SPE Drilling Conference and Exhibition, DC 2018, 2018-March*. https://doi.org/10.2118/189583-MS

Marr, B. (2015, May 26). *Big Data in Big Oil: How Shell Uses Analytics to Drive Business Success*. Forbes. https://www.forbes.com/sites/bernardmarr/2015/05/26/big-data-in-big-oil-how-shell-uses-analytics-to-drive-business-success/?sh=6a2a1c46229e

Mohaghegh, S. D., Gaskari, R., Maysami, M., & Khazaeni, Y. (2014). Data-driven reservoir management of a giant mature oilfield in the middle east. *Proceedings – SPE Annual Technical Conference and Exhibition, 2*, 1056–1079. https://doi.org/10.2118/170660-MS

Mohamed, A., Hamdi, M. S., & Tahar, S. (2015). A Machine Learning Approach for Big Data in Oil and Gas Pipelines. *Proceedings – 2015 International Conference on Future Internet of Things and Cloud, FiCloud 2015 and 2015 International Conference on Open and Big Data, OBD 2015*, 585–590. https://doi.org/10.1109/FICLOUD.2015.54

Mohamed, A., Salah Hamdi, M., Tahar, S., Mohamed, A., Hamdi, M., & Tahar, S. (2017). Data science and big data: An environment of computational intelligence. *Studies in Big Data, 24*. https://doi.org/10.1007/978-3-319-53474-9_9

Mohammadpoor, M., & Torabi, F. (2020). Big data analytics in oil and gas industry: An emerging trend. *Petroleum, 6*(4), 321–328. https://doi.org/10.1016/j.petlm.2018.11.001

MoPNG. (2020). *Digitalization Roadmap for Indian Exploration and Production (E&P) Industry*. http://petroleum.nic.in/sites/default/files/Draft_digitalization_roadmap_final.pdf

Mounir, N., Guo, Y., Panchal, Y., Mohamed, I. M., Abou-Sayed, A., & Abou-Sayed, O. (2018). Integrating big data: Simulation, predictive analytics, real time monitoring, and data warehousing in a single cloud application. *Proceedings of the Annual Offshore Technology Conference, 4*, 2991–3004. https://doi.org/10.4043/28910-ms

Neri, P. (2018). Big data in the digital oilfield requires data transfer standards to perform. *Proceedings of the Annual Offshore Technology Conference, 4*, 2911–2916. https://doi.org/10.4043/28805-ms

Ngosi, R., & Omwenga, J. (2015). Factors contributing to non-productive time in geothermal drilling in Kenya: A case of menengai geothermal project. *American Scientific Research Journal for Engineering, Technology, and Sciences (ASRTJETS), 14*, 16–26. http://asrjetsjournal.org/

Nguyen, T., Gosine, R. G., & Warrian, P. (2020). A systematic review of big data analytics for oil and gas industry 4.0. *IEEE Access, 8*, 61183–61201. https://doi.org/10.1109/ACCESS.2020.2979678

Nicholson, J. P. (2007). *Combined CIPS and DCVG Survey for More Accurate ECDA Data.*

Noshi, C. I., Assem, A. I., & Schubert, J. J. (2018, December 10). The role of big data analytics in exploration and production: A review of benefits and applications. *Society of Petroleum Engineers – SPE International Heavy Oil Conference and Exhibition 2018, HOCE 2018*. https://doi.org/10.2118/193776-MS

NS Energy. (2016, November 16). *GE, Maersk Drilling Start Analytic-Driven Pilot Project to Improve Vessel Productivity*. NS Energy. https://www.nsenergybusiness.com/news/newsge-and-maersk-drilling-to-pilot-marine-digital-transformation-5671100/

Pandey, Y. N., Rastogi, A., Kainkaryam, S., Bhattacharya, S., & Saputelli, L. (2020). *Machine Learning in the Oil and Gas Industry: Including Geosciences, Reservoir Engineering, and Production Engineering with Python*. Apress.

Patel, H., Prajapati, D., Mahida, D., & Shah, M. (2020). Transforming petroleum downstream sector through big data: A holistic review. *Journal of Petroleum Exploration and Production Technology*, *10*(6), 2601–2611. https://doi.org/10.1007/S13202-020-00889-2

Plate, M V, & Ag, C. (2018). SPE-181037-MS Big Data Analytics for Prognostic Foresight New Dimension of Petroleum Asset Management. *SPE Intelligent Energy International Conference and Exhibition*, Aberdeen, United Kingdom.

Putcha, V. B., & Ertekin, T. (2018). A Hybrid Integrated Compositional Reservoir Simulator Coupling Machine Learning and Hard Computing Protocols. *Society of Petroleum Engineers – SPE Kingdom of Saudi Arabia Annual Technical Symposium and Exhibition 2018, SATS 2018*. https://doi.org/10.2118/192368-MS

van Rijmenam, M. (2013). *If Big Data is the New Oil, The Oil and Gas Industry Knows How to Hand*. https://datafloq.com/read/big-data-oil-oil-gas-industry-handle/133

Roy, A., Jayaram, V., & Marfurt, K. J. (2013). Active Learning Algorithms in Seismic Facies Classification. *Society of Exploration Geophysicists International Exposition and 83rd Annual Meeting, SEG 2013: Expanding Geophysical Frontiers*, 1467–1471. https://doi.org/10.1190/segam2013-0769.1

Schlumberger. (2018). *DrillPlan Solution Improves Well Planning Efficiency by More Than 50%*. Schlumberger online.

SEAWANDERER. (2020, January). *Neptune Energy Aims to Find Hydrocarbons 70 per cent Faster with Digital Technology*. https://seawanderer.org/neptune-energy-aims-to-find-hydrocarbons-70-per-cent-faster-with-digital-technology

Srinivasan, S. (Ed.). (2018). *Guide to Big Data Applications* (Vol. 26). Springer International Publishing. https://doi.org/10.1007/978-3-319-53817-4

Technavio. (2015, December 16). *How Oil and Gas is Using Big Data for Better Operations – Technavio*. https://blog.technavio.com/blog/how-oil-and-gas-using-big-data-better-operations

Temizel, C., Purwar, S., Abdullayev, A., Urrutia, K., & Tiwari, A. (2015). Efficient Use of Data Analytics in Optimization of Hydraulic Fracturing in Unconventional Reservoirs. *Society of Petroleum Engineers – Abu Dhabi International Petroleum Exhibition and Conference, ADIPEC 2015*. https://doi.org/10.2118/177549-MS

Udegbe, E., Morgan, E., & Srinivasan, S. (2018). Big Data Analytics for Seismic Fracture Identification, Using Amplitude-Based Statistics. *Proceedings – SPE Annual Technical Conference and Exhibition, 2018-September*. https://doi.org/10.2118/191668-ms

Vega-Gorgojo, G., Fjellheim, R., Roman, D., Akerkar, R., & Waaler, A. (2016). Big data in the oil & gas upstream industry – A case study on the norwegian continental shelf. *Oil Gas European Magazine*, *42*, 67–77.

von Plate, M. (2016). Big data analytics for prognostic foresight. *Society of Petroleum Engineers – SPE Intelligent Energy International Conference and Exhibition*. https://doi.org/10.2118/181037-MS

Wang, T., Li, T., Xia, Y., Zhang, Z., & Jin, S. (2017). Risk assessment and online forewarning of oil & gas storage and transportation facilities based on data mining. *Procedia Computer Science*, *112*, 1945–1953. https://doi.org/10.1016/J.PROCS.2017.08.052

Wang, Z., Zhang, G., Zhou, X., Zhong, X., Li, X., Wang, X., & Julaiti, S. (2012). Reservoir configuration analysis and its significance in oilfield development. *Xinjiang Petroleum Geology*, *33*(1), 61–64. https://en.cnki.com.cn/Article_en/CJFDTotal-XJSD201201016.htm

Yadranjiaghdam, B., Pool, N., & Tabrizi, N. (2017). A Survey on Real-Time Big Data Analytics: Applications and Tools. *Proceedings – 2016 International Conference on Computational Science and Computational Intelligence, CSCI 2016*, 404–409. https://doi.org/10.1109/CSCI.2016.0083

Zaidi, D. (2017, October 26). *Role of Data Analytics in the Oil Industry*. https://towardsdatascience.com/here-is-how-big-data-is-changing-the-oil-industry-13c752e58a5a

Zhang, G., Wang, Z., & Chen, Y. (2018). Deep learning for seismic lithology prediction. *Geophysical Journal International, 215*(2), 1368–1387. https://doi.org/10.1093/gji/ggy344

Zhao, J., Jiang, Y., Li, Y., Zhou, X., & Wang, R. (2018). Modeling fractures and barriers as interfaces for porous flow with extended finite-element method. *Journal of Hydrologic Engineering, 23*(7), 04018024. https://doi.org/10.1061/(asce)he.1943-5584.0001641

Zhao, T. (2018). Seismic facies classification using different deep convolutional neural networks. *SEG International Exposition and Annual Meeting, SEG 2018*, 2046–2050. https://doi.org/10.1190/SEGAM2018-2997085.1

Zhao, T., Jayaram, V., Marfurt, K. J., & Zhou, H. (2014). Lithofacies classification in Barnett Shale using proximal support vector machines. *SEG Technical Program Expanded Abstracts 2014*, 1491–1495. https://doi.org/10.1190/segam2014-1210.1

Zhou, X., Dahi Taleghani, A., & Choi, J. W. (2017). Imaging Three-Dimensional Hydraulic Fractures in Horizontal Wells Using Functionally-Graded Electromagnetic Contrasting Proppants. *SPE/AAPG/SEG Unconventional Resources Technology Conference 2017*, 3575–3582. https://doi.org/10.15530/urtec-2017-2697636

Zhou, X., Tyagi, M., Zhang, G., Yu, H., & Chen, Y. (2019). Data Driven Modeling and Prediction for Reservoir Characterization Using Seismic Attribute Analyses and Big Data Analytics. *Proceedings – SPE Annual Technical Conference and Exhibition, 2019-September*. https://doi.org/10.2118/195856-ms

4 Recent Trends in Natural Gas and City Gas Distribution Network

Namrata Bist, Anirbid Sircar, and Kriti Yadav
Pandit Deendayal Energy University, Gandhinagar, India

CONTENTS

4.1 INTRODUCTION

Natural gas is an environmentally friendly fuel as compared to the other fossil fuels like coal or oil as it burns cleaner, is safer and easier to store and is extremely reliable. With the cost of coal at around 150 USD per ton and cost of oil at around 72 USD per barrel, natural gas costs around 4 USD per MMBtu and is less expensive (business insider, 2021). As depicted in Table 4.1, coal emits about 205 pounds of CO_2 per million Btu, oil (diesel, gasoline etc.) emits about 161 to 157 pounds of CO_2 per million Btu, whereas natural gas emits only 117 pounds of CO_2 per million Btu (EIA, 2021). Due to the cost benefits and emission advantages of natural gas, the industrial applications of natural gas are increasing rapidly.

The use of natural gas for electricity generation and industrial applications is growing rapidly due to environmental benefits like less greenhouse gas emission, usage in re-burning (added to coal or oil fired boilers for reduction of emissions),

DOI: 10.1201/b23013-4

TABLE 4.1
Comparison of CO_2 Emissions from Various Fuels Used in the World

Fuel	CO_2 Emissions in Pounds/MMBtu of Energy Input
Oil	164,000
Coal	208,000
Natural Gas	117,000

reduction in sludge generation due to less amount of SO_2 generation, etc. (Liang et al., 2012). Natural gas is used to fuel around 175,000 automobiles in the United States of America and approximately 23 million automobiles globally. According to a research, the transportation sector produced 138 TMT of CO_2 in 2007–2008, accounting for roughly 7% of total CO_2 emissions in India. With future energy need likely to climb, the input might rise to 346 TMT by 2022 in a business-as-usual scenario, representing a 150% growth. Because of the benefits listed previously, the usage of natural gas as a car fuel is rapidly increasing.

However, for the use of natural gas in powering the vehicles, the gas needs to be compressed and stored in tanks at high pressures. The typical pressures at which the gas is stored ranges between 3000 and 3600 psi. Compressed natural gas (CNG) is one of the growing alternatives to liquid petroleum fuels for transportation of vehicles. There are many advantages to the usage of CNG such as durability of vehicle engines, lower carbon emissions, low costs of the fuel, etc. However development of CNG refuelling infrastructure is vital to the expansion of consumer adoption of CNG vehicles. The present chapter discusses about various forms in which natural gas is used as well as the reasons for its popularity to be used as a transportation fuel. The study also discusses about various fuels under development in the market which poses competition to natural gas as fuel. The study tries to understand about the various smart stations developed to monitor the outflow and usage of natural gas.

4.2 GLOBAL OVERVIEW

In the global scenario of rising crude oil prices and environmental concerns, the role of natural gas has acquired a significant energy requirement. Since 2010, 80% of the growth has arisen in three major areas: In the United States shale, gas revolution has been established; in China where fast growth has been driven by economic expansion and air quality issues; and in the Middle East, natural gas as an economic diversification from oil. According to the BP statistical review of world energy, 2007 natural gas is expected to grow up to 28% by 2025. Natural gas had a spectacular year in 2018, accounting for over half of the growth in global energy demand with a 4.6% increase in consumption. The demand of the energy increases rapidly due to the population and urban growth. The gas demand in Asian countries is mostly from India and China.

Global natural gas production in 2020 has been impacted adversely by the COVID-19 virus. As per reports the global production of natural gas has fell by 2.5% (IEA, 2021). However, there is a steady growth in demand of natural gas which is increasing every year as depicted in Figure 4.1.

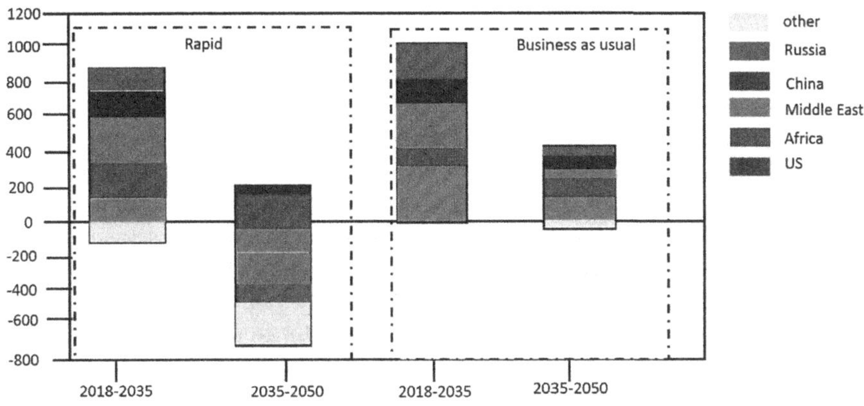

FIGURE 4.1 Outlook for gas is resilient than coal and oil (BP statistical, 2021).

TABLE 4.2
Alternative Fuels in Market

Alternate Fuel	Vehicle Type
Biodiesel	Diesel vehicles
Electricity	Hybrid and plug in vehicle
Ethanol	Flexible fuel vehicles
Hydrogen	Fuel cell vehicles
Natural Gas	Natural gas vehicles
Propane	Propane vehicles

Source: Modified after AFDC, 2021.

During the COVID-19 crisis, although global gas imports had been lowered in the sector of pipeline import, LNG imports had been resilient and have expanded globally by 1.4 Bcm (IEA, 2021). In year 2021, the manufacturing and construction industries are forecast to dominate gas demand growth, with market rising by about 5% as worldwide output and trading levels recover. This expansion is being driven by China, India, as well as other fast developing Asian markets. However, the increase in the renewable capacity uses which and tough price competition from coal impedes the global growth of natural gas.

More than a dozen of alternative fuels are in production or under development for use in alternative fuel vehicles and advanced technology vehicles. As depicted in Table 4.2, fuels such as biodiesel for diesel vehicles, electricity for hybrid and plug in vehicle, ethanol for flexible fuel vehicles, hydrogen for fuel cell vehicles, natural gas for natural gas vehicles and propane for propane vehicles are some of the common examples of alternate fuels.

However, many emerging alternative fuels such as biobutanol, dimethyl ether, methanol, renewable hydrocarbon biofuels, ammonia, etc. are also under development for increasing the energy security, reduction of emissions, improvement of vehicle performance and growth of a country's economy. But the proficient usage of these fuels is bound to take some time. In order to maintain the emission reductions and to improve the energy security, usage of compressed natural gas as the fuel for the vehicles is the most meritorious option amongst the other alternative fuels.

Because of price liberalization, market access liberalization, and the dissociation of transportation from natural gas exports, the retail natural gas market with in United States has altered drastically. Such efforts have aided in the development of a dynamic wholesale market throughout the United States. Whereas the operator sets supply and distribution tariffs on a cost basis, networks have a significant price freedom. As a response, pipeline businesses have found it simpler to respond to shifting market trends. The transformation of the UK natural gas sector from the monopolies to a competitive industry has been accelerated because to structural and regulatory reforms.

Gas transportation services in the United Kingdom are now unbundled, just like they are in the United States at the federal level. National Grid owns and operates the National Transmission System, which transports gas. As in the United States and the United Kingdom, developing competitive natural gas markets and encouraging investment in infrastructure expansion in India may need regular regulatory changes and interventions. Increased regulatory risk and the possibility of political interference, on the other hand, deter natural gas investment. As a result, it is critical to implement structural changes early in the reform process in order to pave the way for expanding markets and competition. Following that, the regulatory structure must continuously be improving to promote market development. Market forces have shown to be significant and effective in the natural gas sector once an acceptable structural and regulatory framework is in place.

CNG has the lowest greenhouse gas emissions among the hydrocarbon fuels making it one of the emerging choices of fuel in vehicle section (Ally and Pryor, 2007). CNG is compressed to less than 1% of the volume and stored at high pressures in cylindrical or spherical vessels. The expansion in the number of CNG powered motor vehicles has corresponded with an expansion in CNG refuelling infrastructure (Frick et al., 2007). CNG fuelling points are operated using either the time fill or fast fill strategy. Time fill strategy allows longer filling times while the fast fill has filling time of less than five minutes (Bang et al., 2014).

4.3 FORMS OF NATURAL GAS AVAILABLE

It is very important to understand the various forms of natural gas available in the world. CNG is often confused with LNG (liquefied natural gas). Natural gas is available in the form of LNG, CBG (compressed biogas), LCNG (liquefied compressed natural gas), HCNG (hydrogen compressed natural gas), etc.

4.3.1 LIQUEFIED NATURAL GAS

Despite the fact that experiments to liquefy natural gas for storage commenced in the 1900s, the world's largest first LNG ship did not carry cargoes from Louisiana to the United Kingdom till 1959, showing the viability of transoceanic LNG transportation.

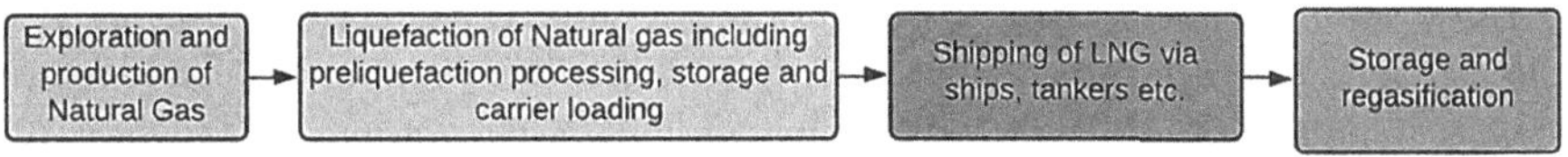

FIGURE 4.2 LNG value chain (IEA, 2021).

Five years later, the United Kingdom started importing Algerian LNG, making Algeria's state-owned oil gas company Sonatrach the world's biggest LNG exporter.

Until 1990, when British North Sea gas became a cheaper option, the United Kingdom continued to import LNG. In 1969, Japan became the first country to import LNG from Alaska, and in the 1970s and 1980s, it pushed itself to the forefront of the worldwide LNG trade with a massive increase in LNG imports. Such supplies aided in the development of natural-gas-fired power production in Japan, reducing pollution and alleviating the effects of the 1973 oil embargo. Japan consumes over than 95% of its natural gas and receives over half of all LNG supplied globally. Figure 4.2 depicts the global LNG industry as a "value chain" with four components: (1) exploration and production, (2) liquefaction (Figure 4.3), (3) shipping and (4) storage and regasification, which provide natural gas for delivery to a variety of "end users." In order to promote economic growth to an LNG project, the cost of a component volume of gas supplied into a piping system has to at least equivalent the

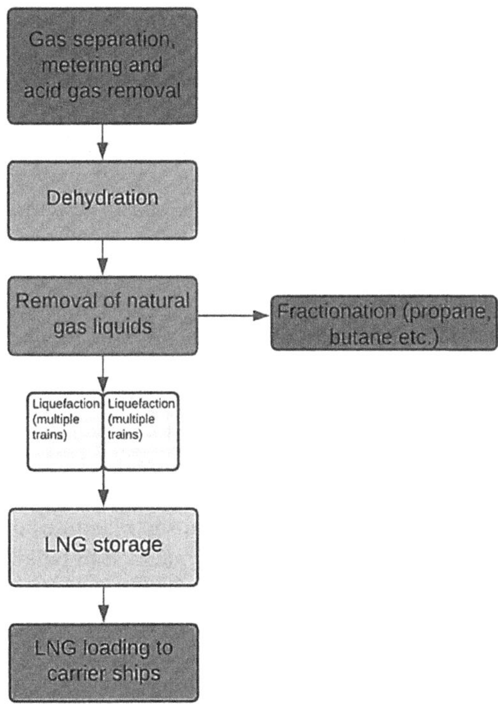

FIGURE 4.3 LNG liquefaction (IEA, 2021).

merged cost to produce, vaporizing, transmitting, storing, and revaporizing the gas, plus the costs necessary to create necessary infrastructure—and a reasonable return to shareholders.

The liquefaction plant typically accounts for the biggest percentage of the overall cost of the LNG value chain, with the production, shipping and regasification components accounting for approximately equal portions of the remainder.

LNG is kept at atmospheric pressure in double-walled, insulated containers with novel, extremely safe and stable designs prior to regasification. The inner tank's walls, which are made of special steel alloys with high nickel content, aluminium and pre-stressed concrete, must withstand cryogenic temperatures. Perlite, a crystalline volcanic material, is mixed with cement concrete and other alloying elements to make LNG storage tanks that are strengthened with steel bars. Those shields shelter the cryogenic container from the elements. Perlite is often used as insulator in the tank's sides. To prevent leakage, many storage facilities have a double-containment mechanism, with both internal and outer walls capable of carrying LNG.

4.3.2 Compressed Bio Gas

The decomposition of waste/bio-mass sources like agriculture residue, cattle dung, solid waste, etc. also produces natural gas which is known as bio gas as shown in Figure 4.4. The bio gas consists of very high methane content. After purification and compression of this gas the gas is called compressed bio gas or CBG. The CBG consists of 90% methane and has exactly the same composition and energy potential (calorific value) as the extracted natural gas from earth's subsurface. Given the abundance of biomass in the world, the CBG has the potential to replace or supplement the CNG in automobile and transportation industry.

Countries such as Germany, Italy, the United Kingdom, France, and Switzerland are encouraging the use of biogas by supporting legal frameworks, education programmes, and technological availability. Biogas is mostly supplied into local natural gas networks and utilized for power production in European nations. Grid injection is the most popular method in European countries, followed by cars that run on biogas (either pure or combined with natural gas), and biogas is also used for heating, either directly or in combination with natural gas. In India, the composition of CBG supplied should meet IS 16087:2016 specifications as shown in Table 4.3. It is suggested that compressed bio-gas facilities be put up primarily by sole proprietors. CBG generated at all these facilities must be transferred via cascades or pipes to Oil Marketing Companies' (OMCs) filling station infrastructures for sale as a green transportation fuel option. To increase investment returns, the entrepreneurs will be able to advertise the additional by-products from such facilities independently, such as bio-manure, carbon dioxide, and so on. It is intended to build 5,000 compressed bio-gas facilities in phases across India until 2024.

The National Policy on Biofuels of 2018 in India emphasizes the active promotion of advanced biofuels, such as CBG. The GOBAR-DHAN (Galvanising Organic Bio-Agro Resources) project was developed by the Indian government to make on farms CBG and compost from cow dung and solid waste. The Ministry of New and Renewable Energy has granted Rs. 4 crore in Central Financial Assistance (CFA) for

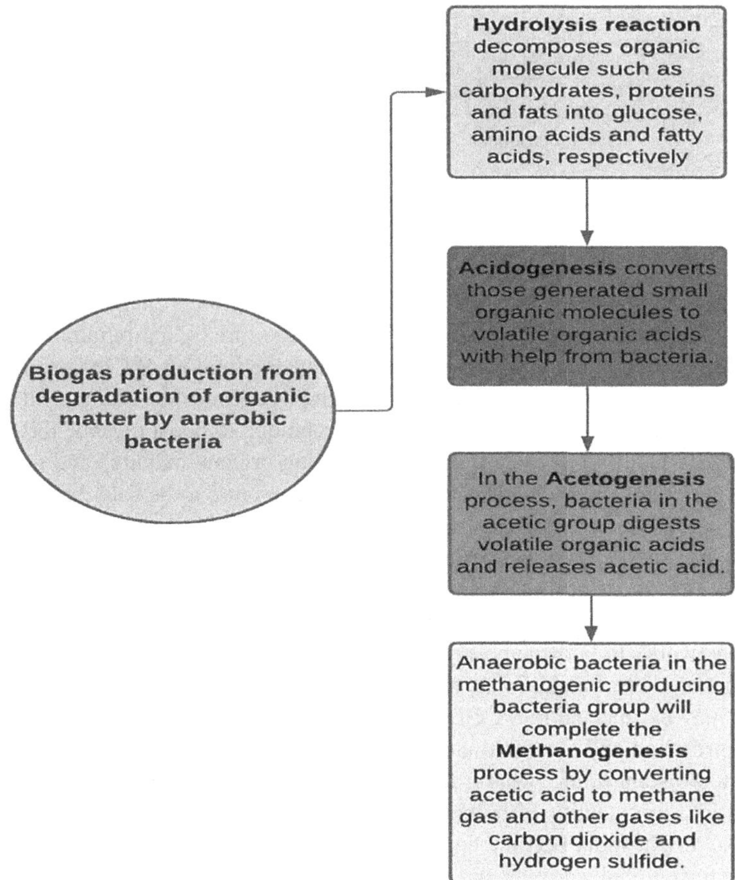

FIGURE 4.4 Production of Biogas by subsequent chemical and biochemical reactions (World Biogas Association, 2019).

TABLE 4.3
Chemical Constituents of CBG

IS 16087: 2016 Standard	
Feature	**Value**
Methane percentage	90%
CO_2 percentage (maximum)	4%
Carbon Dioxide (CO_2) + Nitrogen (N_2) + Oxygen (O_2) percentage maximum	10%
O_2 percentage (maximum)	0.5%
Total sulphur mg/m^3(maximum)	20 mg/m^3
Moisture mg/m^3(maximum)	5 mg/m^3

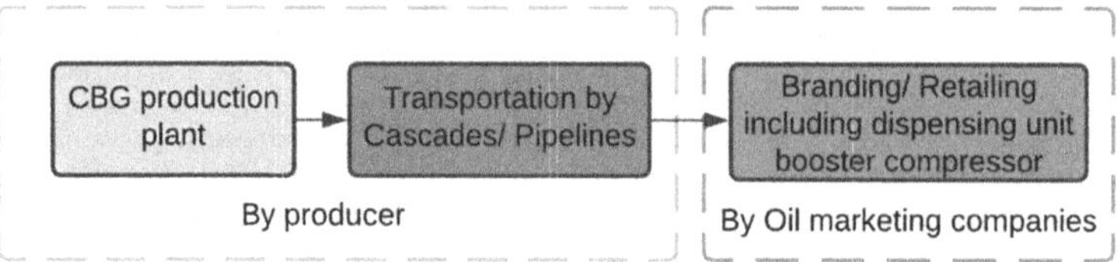

FIGURE 4.5 Envisaged business model for oil marketing companies in India.

each and every 4,800 kg of CBG produced each day from 12,000 cubic metres of biogas produced per day, limited to a total of Rs. 10 crore each programme. Leading Indian banks are collaborating in order to facilitate the funding of CBG projects. The Ministry of Agriculture has included CBG plant bi-products like "fermented organic manure" in the Fertilizer. Implementing this technique to boost revenue through the sale of CBG facility bi-products such "fermenting organic manure" and encourage organic agriculture in India. This enables organic manure to be sold throughout the United States. Member nations have also stepped forward to support CBG efforts. Some states, such as Haryana, Punjab, and Uttar Pradesh, have formed State Level Committees to monitor the development and supervision of the SATAT Programme.

In the future, compressed bio-gas networks might be linked to City Gas Distribution (CGD) networks to boost supply to residential and retail consumers in current and upcoming markets. In addition to retailing from OMC gasoline stations, compressed bio-gas may be injected into CGD pipelines at a later date, allowing for more efficient distribution and increased access to a cleaner and more cost-effective fuel. In the not-too-distant future, what is today considered rubbish will be transformed into electricity! Any quantity of biomass may be converted into biogas or biomanure. This would help India become energy self-sufficient while also contributing to the battle against global warming and climate change.

In India, company called IOT biogas private limited has installed a 2.4 MW power plant. The government of India has taken several initiatives under Sustainable Alternative towards Affordable Transportation (SATAT). In India the envisaged business model for CBG plants is depicted as per Figure 4.5.

As shown in Figure 4.6, the global uses of biomass are bound to increase drastically in the near future.

4.3.3 Liquefied Compressed Natural Gas

LNG has a higher energy density than CNG, making its range more equivalent to that of conventional fuel for long-distance cars. Gas stations can also have liquefied compressed natural gas (LCNG) as the source of CNG for filling the vehicles with the natural gas. The LCNG fuelling pressure should be 200 bar with a 300 bar intermediate storage pressure. The flow range of the station is 500 Nm^3/hr–8,000 Nm^3/hr. A variable frequency drive (VFD) controls the flow regulation, with an average refuelling time of 3–4 minutes at 200 bar. Each dispenser has the capacity to fill 12 cars each hour. The fuelling flow each dispenser of LCNG is 70 Nm^3/hr. LNG storage facilities supply natural gas to LCNG, and cryo pump are used at LCNG facilities

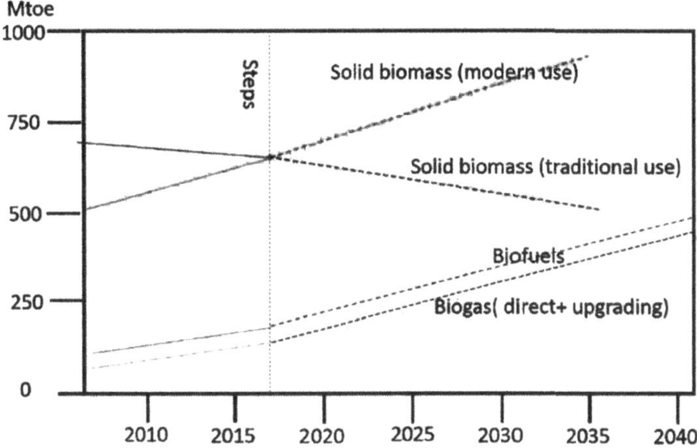

FIGURE 4.6 Breakdown of global biomass demand from 2010 to 2040.

to move LNG from such an insulating holding tank to the automobile through a dispenser. The major parts of an LCNG station are storage vessel, cryogenic pump, vaporizer and dispenser. To make CNG from LNG, the gas is compressed up to 300 bar pressure in a controllable environment, allowing it to be delivered as CNG at the proper pressure (AFDC, 2021). The LNG is pumped into to the evaporator by a high-pressure piston pump on the LCNG skid. This process converts the liquid into gas; the gas is stored in the cylinders and then used to fuel the vehicles. There are two methods for converting LNG to CNG. To synthesize gas, first compress the liquefied gas to a maximum pressure of 200 bar, then heat and evaporate it. Another method for obtaining CNG from LNG is to vaporize or heat the gas directly from the tank and then pass it through an odorizer before compressing it with a conventional gas compressor.

Figure 4.7 shows different liquefied natural gas production and refuelling choices, as well as LCNG stations for mid- and small-scale stations. In a CNG station, a compressor is utilized, but an LCNG station requires a cryogenic LNG pump and LNG storage tanks. High starting expenditure is associated with the LCNG stations due to the high expenses associated with the cryogenic tanks but operation costs are low due to less electricity requirements. Despite the fact that compressor prices are included in some studies, it's impossible to compare the capital and operating expenses of these two types of gas refuelling junctions because liquefier charges are typically not included. Regardless of the techniques employed in the various cost analyses, the LCNG station with larger fuelling capacity appears to offer a better return. Furthermore, by lowering LNG storage demand, LCNG plant construction costs might be lowered. This can be achieved by substituting small-scale, low-cost liquefiers for some of the LNG storage tanks. Despite the fact that CNG-fuelled cars outnumber LNG-fuelled vehicles, infrastructure for high-fuel consumers is more cost-effective (i.e., heavy vehicles). Another advantage of liquefied natural gas is the ability to obtain more constant fuel quality (Landfill Gas Conversion to LNG and LC02, 1999).

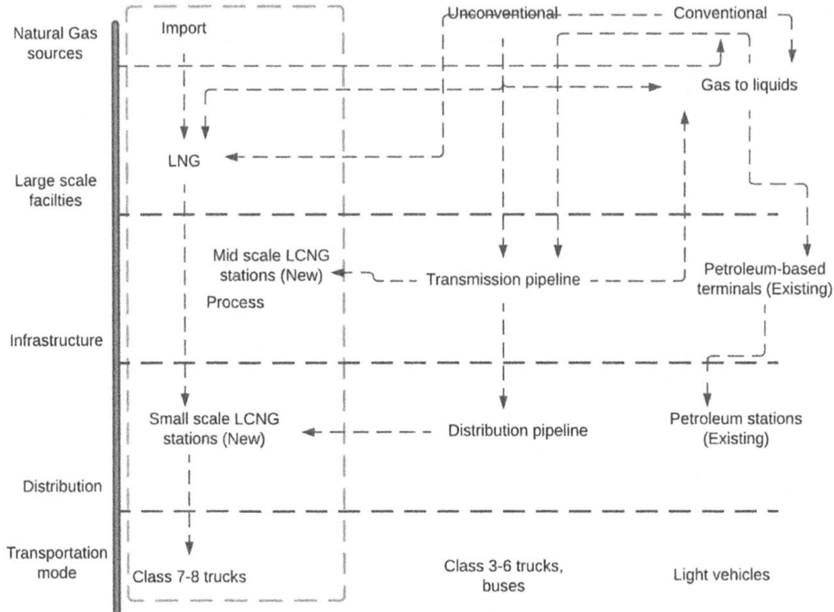

FIGURE 4.7 Liquefied natural gas production and refuelling options. Dotted box denotes the LCNG stations for mid-scale and small-scale stations.

4.3.4 Hydrogen Compressed Natural Gas

Hydrogen can be blended with methane in certain proportions and may act as a transportation fuel for vehicles. Hydrogen compressed natural gas is the name of the mixture (HCNG). CNG emits very little HC and CO_2, as well as very little SOx and PM, rendering it a "clean" fuel. Natural gas also has a major research octane rating of 130, which is an added benefit. This implies that a natural gasoline engine can run at compression ratios with banging than a petrol engine, enhancing thermal performance.

HCNG has properties that are halfway between hydrogen and CNG. HCNG has a number of distinct characteristics that make it remarkably well suited to engine applications in theory. The following are some of the most notable features:

1. Adding hydrogen to a fuel increases the H/C ratio. With a higher H/C ratio, less carbon dioxide is emitted per unit of energy generated, resulting in reduced greenhouse gas emissions.
2. Natural gas has a slow flame, whereas hydrogen has a combustion efficiency that is around eight times faster. As a consequence, natural gas ignition is less consistent than HCNG combustion whenever the excess air ratio is much more than the molar ratio requirement. When using natural gas, there is a risk of incomplete burning (misfire) occurring before acceptable NOx decreases are achieved. If hydrogen is introduced to the fuel, the quantity of charged dilution that can be performed while still maintaining efficient ignition increases (Global potential of biogas, 2019).

3. Because hydrogen does have a low efficiency per unit of volume, the fractional heat capacity of the HCNG combination drops when the quantity of hydrogen in the combination is raised (Nischal and Kumar, 2008).
4. The addition of hydrogen to CNG makes the combination burn more efficiently. As hydrogen levels rise, the mass fraction burnt (MFB) advances. At a given excess air ratio, it gives a shorter combustion time. Because of the high combustion temperature, this would result in increased NOx. As the hydrogen proportion rises, the spark timing should be pushed back. Spark timing is delayed, which minimizes compression effort and lowers combustion temperature. This aids in the reduction of NOx production. The choice of spark timing would be critical in determining the power-to-emissions trade-off (Kavathekar et al., 2007).
5. HCNG combinations of 15–30% prolong the lean working limit, maintaining combustion reaction and lowering HC and CO outputs.
6. Since hydrogen has a laminar burning velocity roughly eight times the same of natural gas, introducing hydrogen to a combination can boost the burning speed, huge benefit such as reduced combustion duration, a higher degree of constant volume ignition, and better apparent thermal performance.
7. By enhancing fossil fuels instead of merely dispossessing them, unique attributes of hydrogen as a combustion stimulant could indeed generate leverage variables much larger than 1; an obvious advantage of the leverage effect is that Reduction process is feasible even though the hydrogen is generated by natural gas with no CO_2 "sequestration" (Ortenzi et al., 2008).

If the existing heavy duty CNG buses are run by HCNG, then HCNG has the potential to meet the toughest diesel engine Euro-V and Euro-VI norms yet to be enforced in future in India as shown in Figure 4.8 (Kavathekar et al., 2007). HCNG can be successfully employed in existing CNG engines with proper modifications.

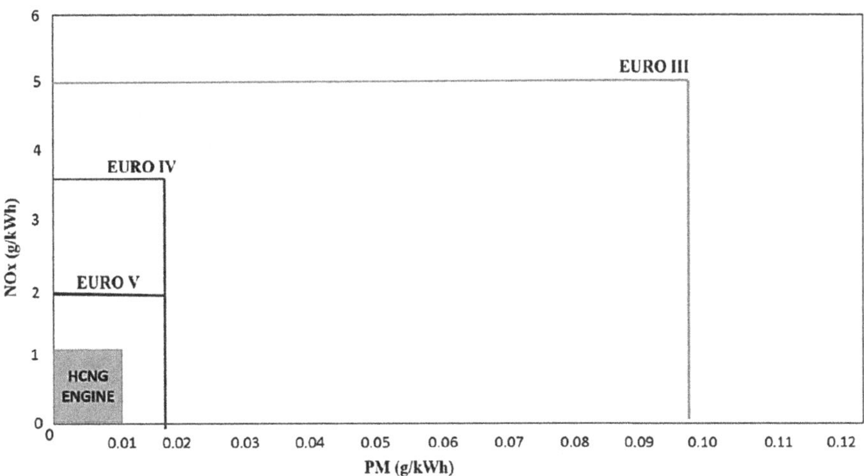

FIGURE 4.8 Emission potential of HCNG engines as compared to other engines of past and present.

4.4 SMART STATIONS (COCO, CODO AND DODO)

Although many competitive fuels are entering into the market as the fuel for transport vehicles, CNG is the most competitive amongst all in terms of costs. There are various types of retail stations for CNG filling based on the operator of the station. Managing a retail chain involves: purchasing land and facilities in high-traffic retail areas, creating and maintaining a facility that is attractive to consumers and keeps them coming back, delivering exceptional customer service across all product and service lines and managing fuel and goods stocks effectively.

Following the issuance of a dealership licence, a corporation may operate under various business models. Companies operating service stations often follow one of three models: This concept is known as a "dealer-owned-dealer-operated model" where the dealer owns and operates the service station ("DODO"). This is the most popular type used in the oil and gas retail industry in Turkey and it is utilized by the majority of stations. The "company-owned-company-operated" model refers to when a distribution business or one of its subsidiaries owns and operates a service station ("COCO"). The "company owned-dealer-operated model" ("CODO"), in which the distribution company or any of its subsidiaries owns the property on which the service station is located and the dealer runs the service station, is another gas and retail business model. In Turkey, the COCO and CODO models are rarely utilized because most dealers are also service station owners in the present petroleum market.

Various types of retail arrangements are as follows.

4.4.1 Company Owned-Company Operated

The company owns the property and employs paid or commissioned workers to run it. Stations managed by a company might be supplied directly or through a branded reseller that provides transportation services. COCO employees are frequently paid in a different way than other company employees. Employees work their way up from pump attendant to dealer, who may oversee and manage a station, under certain multinational marketers' dealer system. They discourage the appointment of open market dealers because such dealers have previously proven untrustworthy. As a consequence, genuine people with a good reputation may start as a pump attendant and work their way up to owning and operating a filling station franchise. People are taught in the process of managing and operating a large marketing gasoline station as they go up the corporate ladder, so that when they are called to a position of responsibility, they are not unfamiliar with the task. Someone who started as a pump attendant would have worked for the firm for 10–15 years and been taught several times before becoming a dealer. As a result, his training as a dealer is limited to managerial and people management, and the firm retains you even after you become a dealer. The dealer does not supply, design, or install CODO stations; instead, the company does everything.

4.4.2 Company Owned-Dealer Operated

The site is owned by the firm; however, it is managed by an independent dealer. The firm rents the dealer the station and land, as well as tanks, pumps, signage and other equipment. The lease terms for the facility are included in the dealer contract, as is the need to purchase fuel at the company's "dealer tank wagon" (DTW) price.

4.4.3 Dealer Owned-Dealer Operated (Open Dealer)

The company owns the tanks, pumps, signage and other equipment, whereas the dealer owns the location. These merchants may be supplied by a refiner or a branded reseller, depending on their supply arrangement. They may move to another source of supplies, including a different brand, when a contract expires. The majority of independent marketers run their retail gasoline outlets under similar company–dealer agreements. Sales to Independent Marketers by gas companies are considered as unbranded product sales under a term contract. Then there are the Dealer Owned Dealer Operated (DODO) stations, which were created by private individuals but later sold to major oil marketing firms. To do so, seek for the largest firm with the finest service structure, a strong brand, and a stock market listing, which will safeguard your investment by gaining investor trust and patronage. Many people contact these firms seeking a franchise and a connection with the finest global brand, either by coincidence or by experience, or those who have done their homework well. They provide a window for you in terms of advertisements, better supplier discounts and so on because you invest money, allowing you to obtain a larger return on your investment and a higher profit margin. For DODO stations, the corporations do independent research to see how many of the company's requirements have been satisfied, and if the parameter passes the company's approval, the station is then supplied with stocks and branded.

4.5 SMART STATIONS

Smart CNG stations are rapidly emerging, including features like as smart dispensers, smart meters and smart card systems. Smart CNG dispensers are used to dispense CNG at filling stations for cars on the road. Dispensers in the modern era are small systems. One-sided or double-sided smart CNG dispensers with one to four hoses are available. The CNG dispensers come with a high-quality pressure system, a Coriolis mass flow metre and dependable electronic calculator technology. The CNG dispensers may be used in two modes: manual (off-line) and automatic (where the gasoline dispenser is connected to the kiosk control system and data on volume, total and pricing is sent to the POS). LCD screens or multimedia displays with controlled LED lighting are used in the systems. For each dispensing nozzle, the system includes electronic and electromechanical totalizers. An infrared remote controller for the dispenser might be included in the dispenser configuration. Electromagnetic valves are used to regulate pressure sections in a sequential manner. The portion of the dispenser that controls the amount/volume might feature a four- or twelve-key keypad. The dispensers are made to withstand pressures of up to 200 bar and cold temperatures. Intelligent switching methods are used in the pressure portions. Signal bacons and diagnostic lines assist the dispensers.

Many advanced CNG dispensers ensure fast, safe and accurate filling. Their modularity allows for installation next to liquid fuel dispensers and integration with the station management and monitoring systems. The modern dispensers allow instant flow capacities up to 50 kilograms per minute. The refilling mechanisms consist of mass meters installed at every hose for a priority refilling system.

The dispensers consist of excess flow shut-down system. In case of any breakage the break-away valves are also installed in the system. The Emergency shut-down system is installed at the dispensers to help in the case of sudden emergencies. The present smart dispensers are also aided with automatic compensation systems depending on the ambient temperatures while deciding the filling pressure. The display screens are of many types; most attractive ones are liquid quartz display which indicates important parameters like price of gas per meter cubes, total price of sale and total volume etc.

The smart stations of today are also able to store previous sales, accumulated volumes and total sales information. These stations are able to have serial communication with the station management. The smart dispensers are designed to be used for any fluid such as oil, air or water.

The CNG stations are also equipped for handling smart cards. The smart cards can be with contact cards or contactless cards. The machine has amenities like automatic card lock and eject function. These devices are clog-proof, dust-proof and are environmentally adaptable. The interface of the devices can be customized according to the requirement of users. The design is also shock proof to avoid card ejection by external force.

New smart stations consist of amenities like Kiosk terminals, Fuel Dispensers, CNG Dispensers, Parking Systems, Ticketing Machines and Vending Machines, etc. The stations consist of power failure protection systems.

Along with the technical advancements, smart stations of today consist of mini marts and coffee shops. These experiences make the customer attracted towards the station. In long run, these amenities would attract the middle class and upper middle class as well as upper class population to shift to CNG vehicles.

4.6 CONCLUSION

CNG is natural gas that has been compressed to a pressure of 200–250 bars and is in a gaseous condition. It is utilized as a car fuel. Compression, storage, handling and transportation through cascades are all part of the CNG station's activities, which may be necessary for marketing, selling and dispensing. The CNG sector is one of the rapidly growing sectors in the world. Many factors are responsible for the growth of this sector and are known as growth drivers or demand drivers of CNG. Rising cost of coal, fast-growing economy, rising price of crude oil, environmental concerns, availability of affordable natural gas and enabling policy framework, infrastructure development, etc., are some of the growth drivers of the CNG. The creation of a natural gas transportation market would encourage transparent and equitable pricing. A switch from the current zonal tariff system to a suitable tariff system, such as the "Entry Exit system" or a "hybrid system," that incentivizes competition in the market while also providing services to customers at a reasonable tariff, particularly in geographically unfavourable locations, with appropriate customization taking into consideration multi-ownership, tax systems and other variables. In the Indian situation, significant infrastructure investment in LNG import and domestic gas transportation is anticipated for the next 5–6 years. CNG vehicle technology is well-established and easily obtainable for all types of road transport vehicles throughout the world. CNG has a number of benefits over diesel and petrol, notably lower emissions and costs, as well as making nations more energy independent by lowering their reliance on oil. Natural gas as a transport fuel has the ability to improve air quality in urban areas, reduce adverse health impacts and lower the cost to society of air pollutants.

ABBREVIATIONS

CBG	Compressed Biogas
CFA	Central Financial Assistance
CGD	City Gas Distribution
CNG	Compressed Natural Gas
COCO	Company Owned-Company Operated
CODO	Company Owned-Dealer Operated
DODO	Dealer Owned-Dealer Operated
EIA	Energy Information Administration
GOBAR-DHAN	Galvanising Organic Bio-Agro Resources
HCNG	Hydrogen Compressed Gas
IEA	International Energy Agency
LCNG	Liquefied Compressed Natural Gas
LNG	Liquefied Natural Gas
SATAT	Sustainable Alternative towards Affordable Transportation

REFERENCES

Ally, J., and Pryor, T. 2007. Life-cycle assessment of diesel, natural gas and hydrogen fuel cell bus transportation systems. *Journal of Power Sources*, 170(2), 401–411.

Alternative Fuels Data Center: Compressed Natural Gas Fueling Stations, 2021. Retrieved 11 September 2021, from https://afdc.energy.gov/fuels/natural_gas_cng_stations.html

Bang, H. J., Stockar, S., Muratori, M., and Rizzoni, G. 2014. Modeling and analysis of a CNG residential refueling system. In *Dynamic Systems and Control Conference* (Vol. 46209, p. V003T37A005). American Society of Mechanical Engineers.

BP, 2021. Retrieved 11 September 2021, from https://www.bp.com/en/global/corporate/energy-economics/energy-outlook/demand-by-fuel/natural-gas

Energy Information Administration (EIA). 2021. Retrieved 11 September 2021, from https://www.eia.gov/tools/faqs/faq.php?id=73&t=11

Frick, M., Axhausen, K. W., Carle, G., and Wokaun, A. 2007. Optimization of the distribution of compressed natural gas (CNG) refueling stations: Swiss case studies. *Transportation Research Part D: Transport and Environment*, 12(1), 10–22.

Global potential of biogas, 2019. Retrieved 11 September 2021, from https://www.worldbiogasassociation.org/wp-content/uploads/2019/09/WBA-globalreport-56ppa4_digital-Sept-2019.pdf

International energy agency (IEA), 2021. Retrieved 11 September 2021, from https://www.iea.org/reports/natural-gas-information-overview

Kavathekar, K. P., Rairikar, S. D., and Thipse, S. S. 2007. Development of a CNG injection engine compliant to Euro-IV norms and development strategy for HCNG operation (No. 2007-26-029). SAE Technical Paper.

Liang, F. Y., Ryvak, M., Sayeed, S., and Zhao, N. 2012. The role of natural gas as a primary fuel in the near future, including comparisons of acquisition, transmission and waste handling costs of as with competitive alternatives. *Chemistry Central Journal*, 6(1), 1–24.

Nischal, T. S., and Kumar, A. 2008. Natural Gas Scenario In India-The Recent Upswings, Concerns And The Way Forward. In *SPE Asia Pacific Oil and Gas Conference and Exhibition*. OnePetro.

Ortenzi, F., Chiesa, M., Scarcelli, R., and Pede, G. 2008. Experimental tests of blends of hydrogen and natural gas in light-duty vehicles. *International Journal of Hydrogen Energy*, 33(12), 3225–3229.

World Biogas Association. 2019. Retrieved May 5, 2022, from https://www.worldbiogasassociation.org/global-potential-of-biogas/.

5 Solar Power Forecasting Using Long Short-Term Memory Algorithm in Tamil Nadu State

N. V. Haritha
Meenakshi Sundararajan Engineering College, Chennai, India

Jose Anand
KCG College of Technology, Chennai, India

CONTENTS

DOI: 10.1201/b23013-5

5.1 INTRODUCTION

On July 1, 1957, the Tamil Nadu Electricity Board (TNEB) got into the field of power producer and has endured into an electricity company and supply provider up to date. In this duration, the state government has prolonged the electric community service to the entire villages as well as cities across the Tamil Nadu state. Then 53 years of adventure and on November 1, 2010 the name got updated itself from TNEB Ltd to Tamil Nadu Generation and Distribution Corporation (TANGEDCO) Ltd [1]. To fulfill the power desires of the country, TNEB has a complete hooked-up potential of 10237.41 MW that incorporates important percentage and autonomous power generation units. Moreover, the Tamil Nadu state has connections in renewable energy re-assets like windmill energy, biomass energy and cogeneration as much as 7,303 MW. Until now, the complete hooked up potential in the state is 17,540 MW. The energy produced is transferred and allotted to all of the purchasers inside the Tamil Nadu state at suitable voltage stages. TANGEDCO takes a client base of approximately 22,34,4000 customers. Hundred percentage rural electrification is made. The per-capita intake of Tamil Nadu state is 1,040 units. To obtain the aim of electrical connection to all households, the state has released the Rajiv Gandhi Grameen Vidyutikaran Yojana (RGGVY) system, in which the grid connection isn't viable or now no longer cost operative, decentralized dispersed generation is used [1]. Solar photovoltaic forte productions are growing in maximum of the nations both through grid-connection or stand-on my own networks. Solar photovoltaic (PV) addition calls for the functionality of managing the indecision and variations in electricity. Precise estimation offers the grid machinists and power machine engineers with considerable records to lay out a finest sun photovoltaic plant in addition to dealing with the power of call for and supply [2]. Accurate forecasting can be obtained by the direction of solar radiation and temperature on the photovoltaic cell [3].

Automated Meter Reading (AMR), smart meters and Advanced Metering Infrastructure (AMI) have created a number of hobby initiation in the beyond few years. One of the riding forces has been the directives on power end-use and power services, collectively with marketplace liberalization and a well-known fashion for power saving and environmental difficulty via way of means of consumers. This has created a call for interoperable answers for meter reading. In India, virtual microcontroller/Application-Specific Integrated Circuit (ASIC) primarily used to obtain full power in electricity distribution monitoring. These meter readings and the total energy consumption are used to find the total load profile for constant no. of days. Energy meters stores month-to-month power intake statistics which receives downloaded to the Meter Reading Instrument and in addition to the Base Computer Software (BCS). The intake statistics from BCS receives imported to the billing software program and then payments are generated and dispatch to the consumer. This entire technique takes an excessive amount of time as meters are of various make/producer and BCS is likewise producer unique at the same time as doing AMR or Remote Meter Reading (RMR) using handheld devices. Hence, electricity distribution application calls for a system/merchandise via way of means of which a couple of producers make meter having exceptional conversation protocol, the conversation media and database ought to be interoperable or there ought to be a few open conversation requirements for energy metering accompanied via way of means of all producers. Efforts have been carried out in each of the instructions to make current meters usable for AMR or AMI in addition to new requirements advanced for brand new energy meters. This work presents an accurate technique that is industrialized for solar energy prediction in a current research by the authors. The correctness of the prediction of solar energy is better compared with the obtained real-time results.

The remaining article is prepared as follows. Section 5.2 reviews about the related literature in the solar energy prediction, the system architecture and design description is explained in Section 5.3. Section 5.4 details the LSTM network implementation and Section 5.5 describes the data flow explanation. Section 5.6 carries out the results and discussions and finally conclusion and future scope is furnished in Section 5.7.

5.2 RELATED WORKS

This phase describes the numerous literature in the associated fields. Multiple techniques to forecast solar photovoltaic (PV) power era in view of trends and average overall performance are noted and labeled the forecasting into time-based statistical techniques, physical techniques, and collaborative techniques [2]. Assessment of universal solar energy, mobile temperature and solar energy era prediction are furnished with a relative assessment by the usage of the actual statistics, this is accumulated from a house located in Anadolu University İki Eylül campus in Eskişehir; accordingly, hourly international solar energy cost on horizontal ground are anticipated from the parameters calculated every day, international solar energy costs with the resource of the usage of eleven particular models. The greatest precise replicas are encouraged to be approved in any site that has comparable climatic circumstances with the identified city [3]. The electricity control version offers easy electricity that consists of many re-assets which are linked to shape small groups. The forecasting techniques offer excessive performance to the easy electricity and vital statistics approximately the simple concepts and requirements of PV energy forecasting via way of means of pointing out several researches performed at the PV energy forecasting subject matter specially in the short-time period horizon. This technique is used for PV energy forecasting in conjunction with a cautious look at various time and spatial horizons [4]. For people and small groups energy forecasting approach primarily based totally on gadget mastering, picture processing, and acoustic class strategies are proposed which will increase the manufacturing of sun energy on the customer degree and require automatic forecasting structures to environmental effects, cost, and financial loses for houses and groups that harvest and devour energy. In the electricity marketplace, prosumers, need new synthetic neural community overall presentation tuning strategies to make correct predictions [5].

A technique to forecasting one-day-in advance PV energy technology without the usage of numerical climate prediction statistics, a closed-loop non-linear autoregressive synthetic neural community version with the handiest historic generated photovoltaic energy statistics as input and the complete forecasting machine runs a hazard of failing. A correct cloud picture forecasting could be very vital for sky picture primarily based totally on sun energy forecasting to offer a method which can tune the cloud deformation method after which forecast cloud form and role in a destiny sky picture. The pixel value obtained from sky pix is recognized by the usage of Otsu's approach. A scientific explanation approach of cloud position, formation, and deformation is recommended, so that the important thing records of cloud and its deformation method may be removed from unique pix after which it is digitized. Genetic Algorithm (GA) [6, 7] is implemented to enhance the cloud deformation method in steps with digitized historic cloud records. In the end, the cloud form and role in a destiny picture has estimated the usage of linear extrapolation. The viability and efficiency of the planned approach are demonstrated via way of means of simulation [8]. Lack of predictability of sun energy stays one primary difficulty to the advent of large-scale sun electricity manufacturing. One of the maximum difficult elements of sun prediction is the prerequisite for terribly short-time period prediction because of cloud motion, ambient temperature variant and humidity levels, which bring about speedy rise up and slope down rates. A three-layered feed ahead with lower back propagation version is planned to obtain the neural community regulations [9]. A hybrid version thinking about physical means of Numerical Weather Prediction (NWP) statistics group and generate particular size of irradiation in conjunction with Artificial Neural Networks (ANN) and different statistical strategies for bias correction of irradiance and AC energy [10, 11]. Economic advantages are received from the usage of probabilistic forecasts, for the reason that they enhance the selection-making competencies inside power markets [12]. Probabilistic sun energy forecasting the usage of deep mastering methodologies is used for energy prediction [13]. Because of the progressed deep mastering strategies, reputation is received and they are regularly used in phrases of accuracy as compared to different statistical forecasting strategies [14]. The manner of forecasting processes is made by means of beyond researchers, which includes regressive techniques and gadget mastering strategies; therefore, researchers discover that different statistics re-assets as satellite television for personal computer sensing or faraway sensing, NWP [15]. A fuzzy primarily based totally sun irrigation machine is mentioned in [16]. A green short-time period forecasting of sun energy manufacturing variational car encoder version using deep mastering to enhance forecasting accuracy in period collection showing and bendy nonlinear calculation, that satisfies the overall presentation of deep mastering strategies to predict sun energy and always plays higher than different techniques [17]. A hybrid version of assist vector regression and progressed adaptive genetic set of rules is applied for an hourly power call for forecasting on the hybrid technique outperformed the conventional feed-ahead neural networks, the intense mastering gadget version, and the Support Vector Regression (SVR) version [18].

A univariate approach is recommended for more than one step in advance of sun energy forecasting by means of integrating a statistics re-sampling technique with gadget mastering processes with gadget mastering algorithms which includes neural networks, SVR, random forecast and more than one linear regression are implemented to for computing more than one step in advance predictions that's designed for handiest invariant time collection statistics [19, 20]. An analog approach for day-in advance nearby PV energy forecasting is delivered primarily based totally on climatological statistics, and sun time and earth declination angle, and this approach showed higher day-in advance nearby energy forecasting as compared to the patience version, machine marketing consultant version and SVM version [21]. A technique for predicting PV and wind energy the usage of the highly-ordered multivariate Markov Chain version is mentioned and this method reflects the time-adaptive invariant correlation among the wind and PV output energy to acquire the 15 minutes in advance prediction and the statement c program language period of the closing stated examples are covered to comply with the sample of PV/wind energy variations [22]. A forecast method uniting the gradient boosting set of rules with characteristic production strategies is planned to find records from a grid of arithmetic climate predictions by the usage of each sun and wind statistics, that implies the suitable functions extracted from the uncooked NWP may want to enhance the forecasting [23] using cloud computing [24]. A generalized ensemble version integrating deep mastering version to generate correct sun energy forecasts for twenty-one sun PV centers applied in Germany is studied and on this climate parameter for sun energy technology is chosen as a characteristic method, and the climate class method is used to group the dataset and for every group, a separate collective set of rules is allocated [25]. A two-step version that links surprise climate variables with introduced climate forecasting is applied and the technique improves a base technique via way of means of huge margins no matter sorts of implemented gadget mastering algorithms [26].

5.3 SYSTEM ARCHITECTURE AND DESCRIPTION

5.3.1 Automatic Meter Data Acquisition

Automatic meter data acquisition has appeared because of the analysis of an application meter through a method that doesn't require complete admission to or visible inspection of the meter. A standard meter statistics acquisition machine has numerous fundamental components: meter, Meter Interface Unit (MIU), a communications community and a host computer. Normally, inside an AMR machine, the meter statistics is surpassed from the meter to MIU, which can be outside to the meter or included inside the frame of the meter. In addition to the meter statistics, different pertinent facts can be saved inside the MIU, which includes any tamper or alarm conditions. This tool conveys the interface between the meter and the communication network as illustrated in Figure 5.1.

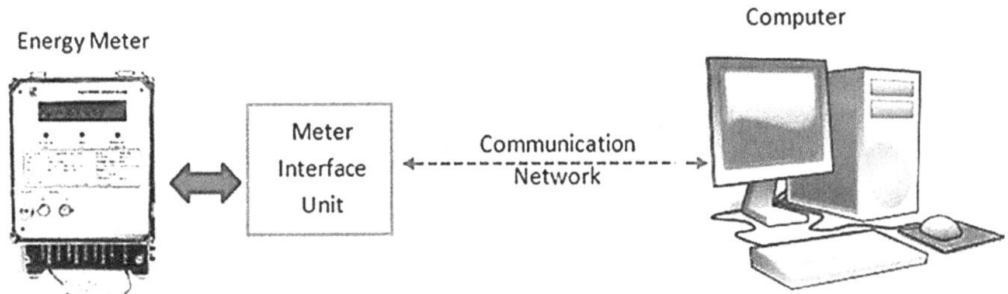

FIGURE 5.1 Automatic meter data acquisition.

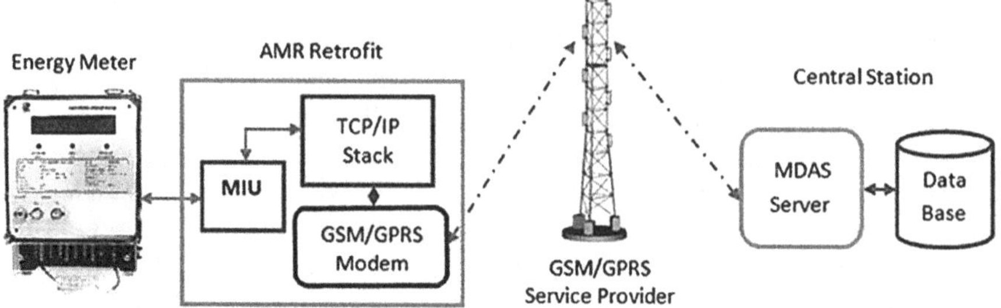

FIGURE 5.2 GSM communication for AMR.

5.3.2 GSM Module Communication

In Global System for Mobile (GSM) communication a present cell community is applied for information transportation and calls for no extra gadget or software, ensuring a sizeable financial savings in each time and capital. The GSM/General Packet Radio Service (GPRS) cell generation makes use of an encryption method to save an outdoor supply from receiving the transmitted information therefore to make certain safety of the meter information. The cell community offers complete two-manner communications, permitting scheduled reads, call for reads, alarm and occasion reporting, electricity outage reporting and electricity healing reporting. Figure 5.2 indicates the GSM conversation module.

5.3.3 Meter Interface Unit

The MIU incorporates the foremost additives together with a modem module. This additionally has an electricity line coupler for 230 V. The microcontroller with outside Electrically Erasable Programmable Read-Only Memory (EEPROM) has a conversation port. A specific silicon ID is fused on this EEPROM. Through hand-held tool meter number, preliminary meter analyzing and meter constants are programmed. This unit also has a rectifier circuit which is used for feeding every submodule.

5.3.4 ZigBee Based AMR System

ZigBee is a brand new wireless networking generation that uses small, low-power virtual radios primarily based totally on the 802.15.4 IEEE standard preferred for Wireless Personal Area Networks (WPANs). It has lately been evolved for far-off metering software and that made ZigBee the state-of-the-art opportunity for AMR. A ZigBee-primarily based totally AMR gadget acquires power intake statistics from a meter automatically. Data are transmitted through a wireless mesh community infrastructure to a server, which may be placed distance away. Each ZigBee community is diagnosed with the aid of using 16-bit community ID.

5.3.5 GSM-ZigBee Coordinator

GZC coordinates the Wi-Fi power meter community nodes and acts as a bridge among GSM conversation and 2.4 GHz wireless ZigBee conversation. It is liable for deciding on the 2.4 GHz wireless channel and Network ID. The GZC begins off a brand new community on strengthening up. Once it has commenced a community, the GZC can permit power meter routers and power meter nodes to enroll in the ZigBee community. The GZC can transmit and get hold of Radio Frequency (RF) statistics transmissions, and it may help in routing statistics through the mesh community. It acts as a bridge among faraway server and strengthens the meters for all communications.

5.3.6 Characteristics of Solar Irradiation

The following are the characteristics of solar irradiation.

i. Data
ii. Component
iii. Global Horizontal Irradiation (GHI)
iv. Diffuse Horizontal Irradiation (DHI)
v. Direct Normal Irradiation (DNI)

The sources for the solar irradiation controlling are:

i. Ground Measurements
ii. Satellite Data

***Measuring Sun Radiation*:** The sun radiation on the earth surface combines Direct Normal Irradiation (DNI) and Diffuse Horizontal Irradiation (DHI). Both are linked in the formula for Global Horizontal Irradiation GHI). The sun radiation is measured by the following.

$$\text{GHI} = \text{DHI} + \text{DNI} \bullet \cos(\theta)$$

Where, θ = solar zenithal angle

$$\text{Sunny day} = 100\%\ \text{GHI with } 20\%\ \text{DHI and } 80\%\ \text{DNI} \bullet \cos(\theta)$$

***Global Horizontal Irradiation (GHI)*:** The overall quantity of energy obtained from above a floor horizontal to the ground. This price consists of each DNI and DHI. It is the factor of GHI which does now no longer come from the beam of the sun.

***Direct Normal Irradiation (DNI)*:** DNI is the amount of sun energy obtained in line with unit region through a surface. This is usually held every day perpendicular to the rays that are available immediately from the path of the solar at its modern role in the sky. Here the attitude of radiation may be 90 degrees.

Diffuse Horizontal Irradiation (DHI): DHI is the quantity of energy obtained according to single place via way of means of a surface (no situation to some color or shadow) that doesn't reach on a right away direction from the sun, however has been dispersed in the form of molecules and debris in the ecosystem and is derived similarly from all directions.

5.3.7 Pyrometer

Pyrometer measures global irradiance in which the quantity of solar power per unit area per unit time occurrence on the surface of precise location originating from a semicircular arena of opinion, represented by $Eg\downarrow$. The universal irradiance shown in Figure 5.3 comprises straight sunlight and diffused sunlight. The influence from straight sunlight gives E • cos (θ),

where

θ = angle between normal surface and sun position in the sky
E = maximum amount of direct sunlight

Thus the universal irradiance is

$$\text{Eg}\downarrow = \text{E}\bullet\cos(\theta) + \text{Ed}$$

5.3.8 Modules in Solar Irradiance Forecasting

EDA and Cleaning Module: Exploratory Data Analysis is a technique used to read the recorded units and to summarize their major characteristics, for the regular use of statistical snapshots etc. Cleaning procedure is like disposing of undesirable records from the Comma Separated Values (CSV) report and keeping the records for destiny purposes. In this text, we eliminated undesirable records columns or rows from the CSV report and created the date, time column that matches with Chennai Time zone. We calculated the Mean, Standard deviation, max, min, etc. by visualizing the

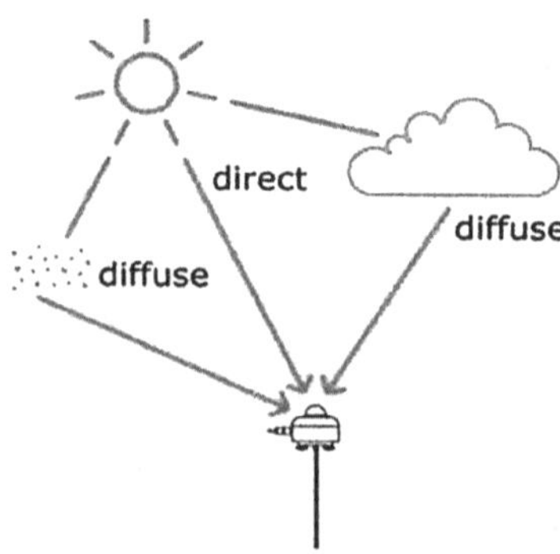

FIGURE 5.3 Global horizontal irradiance measurement.

characteristic correlation and GHI, DNI, DHI over the term and stored the CSV report into pickle format.

***Modeling and prediction Module*:** In this module, we examined the wiped clean pickle document and described the label in the Data body. Resampled the facts body to get quicker consequences and after checking the null values we eliminated that. By scaling the facts and creating lags for proposing the facts body, we split the CSV facts into components: one is educate and every other is test. Train facts are like 67% of normal facts to educate the Long Short-Term Memory (LSTM) Recurrent Neural Network (RNN) version and rest of the facts are checking out. Finally, we matched the fashions and were given the consequences from the version and visualized the consequences using matplot lib and we saved the consequences in an Excel document for future use.

5.3.9 System Implementation

The PV gadget is needed for the conversion of sun power to electricity. However, having the environmentally pleasant function cannot assure the popularity of PV as an opportunity to traditional power sources. The techno-reasonably priced have a look at for viability of any PV gadget calls for appropriately estimating the power yield with the right mathematical model. Different fashions have been proposed for the prediction of PV module performance. However, those fashions ordinarily have a complex shape requiring certain understanding of the parameters which can be generally unavailable in the manufacturer's facts sheet. Therefore, such fashions aren't appropriate for output energy calculation. It is the ambient temperature and sun irradiance meteorological facts that govern the output of a PV gadget. Therefore, dependable temperature and radiation facts need to be quite simply having layout of a techno-economically feasible PV gadget. Because of the problems in set up, calibration, upkeep and excessive fee for dimension of those facts, they may be both now no longer to be had or best in part to be had on the setup site. Hence, the call for exists for the improvement of opportunity approaches to expect them. Solar irradiance is likewise often variable; due to this, capable techniques of forecasting are required for allowing more penetration of sun energy. Influence of location, climate and different meteorological elements additionally make forecasting sun irradiance a difficult task. Therefore, for a hit integration of sun power with conventional era supplies, the capacity of appropriate forecasting sun irradiance is essential.

Data Collection: Data are a group of facts, figures, objects, symbols, and activities accumulated from exceptional re-assets. Organizations gather statistics to make higher decisions. Without statistics, it would be hard for agencies to make suitable decisions, and so statistics is amassed at numerous factors in time from exceptional audiences. For instance, earlier than launching a brand new product, an agency desires to gather statistics on product call for, patron preferences, competitors, etc. In case statistics isn't collected19 beforehand, the agency's newly released product might also additionally result in failure for major reasons, inclusive of much less call for and lack of ability to satisfy patron desires.

Time Series: This refers to a sequential order of values of a variable, referred to as a trend, at identical time intervals. Using patterns, an agency can be expecting the call for its services and products for the projected time.

Data Cleaning: Data cleansing is the technique of solving or eliminating incorrect, corrupted, incorrectly formatted, duplicate, or incomplete statistics inside a dataset. When combining a couple of statistics re-assets, there are numerous possibilities for statistics to be duplicated or mislabeled. If statistics re incorrect, results and algorithms are unreliable, despite the fact that they will appearance correct. There isn't any one absolute manner to prescribe the precise steps in the statistics cleansing technique due to the fact the methods will range from dataset to dataset.

Data Preprocessing: Preprocessing denotes back to the alterations carried out to our statistics former than nourishing it to the set of rules. This technique is used to alter the uncooked statistics accurately into a relaxed statistics set. In other way, on every occasion the statistics is accumulated from exceptional data that give unclear information is not viable for the examination.

Need of Data Preprocessing: For accomplishing higher effects from the carried out version in Deep Learning (DL) initiatives the layout of the statistics is in right means. Approximately certain DL version desires facts in a certain layout, for instance, LSTM RNN set of rules does now no longer guide null values, consequently to execute LSTM RNN null values essential to be controlled from the unique unprepared dataset. Additional component is the statistical set ought to be arranged in any such way that multiple ML and DL algorithms are finished in a sole information set, and excellent out of them is selected.

Rescale Data: As statistics is constituted of characteristics with various rules, numerous gadget studying algorithms will advantage from rescaling the qualities to all having the equal levels. This is beneficial for optimization procedures used in the middle of gadget studying procedures like gradient ancestry. It is likewise beneficial for procedures that measures inputs like regression and neural community and processes that creates distance evaluation like K-Nearest Neighbors. This is modified to our statistics using scikit which is used to examine the use of the Min Max Scaler class.

Model Building: DL works with—fashions and—algorithms, and each play a vital function in deep studying in which the set of rules tells approximately the technique and version is constructed via way of means of following the ones rules. A version in deep studying not anything however a characteristic used to take a few certain inputs, and to carry out certain operation that is instructed by means of algorithms to its excellent at the given input, and offers an appropriate output.

Model in DL: The version is depending on elements inclusive of functions selection, tuning parameters, price capabilities together with the set of rules, the version simply now no longer completely depending on algorithms. Model is the end result of a set of rules while we put into effect the set of rules with the code while we teach the algorithms with the actual statistics. The version is used to expect the destiny end result which is discovered by means of the set of rules implemented with small data.

$$\text{Model} = \text{Data} + \text{Algorithm}$$

LSTM RNN: In this, we evolved an LSTM for the human interest popularity dataset. LSTM community fashions are a form of recurrent neural community which might be capable of examine and keep in mind over lengthy sequences of entering statistics. They are meant to be used with statistics this is constituted of lengthy sequences of statistics, as much as 200-to-400-time steps. They can be an amazing match for this problem. The version can guide a couple of parallel sequences of enter statistics, inclusive of every axis of the accelerometer and gyroscope statistics. The version learns to extract functions from sequences of observations and the way to map the inner functions to exceptional interest types. The advantage of the use of LSTMs for series type is they can examine from the uncooked time collection statistics directly, and in turn which will not require area knowledge to manually engineer input functions. The version can examine an inner illustration of the time collection statistics and preferably attain similar overall performance to match on a model of the dataset with engineered functions. This section is categorized into three sections; they are as follows:

i. Load Data
ii. Fit and Evaluate Model
iii. Summarize Results

5.4 LONG SHORT-TERM MEMORY NETWORKS

LSTM is type of RNN, skilled to learn long-term requirements developed by Hochreiter are Schmidhuber on 1997, and were sophisticated and promoted by numerous other people in various works. These are tremendously applicable for huge diversity of glitches and are extensively utilized. These are openly intended to evade the long-term reliance problems. Memorizing information for extensive phases of time is nearly their evasion conducted, not somewhat they brawl to learn. All recurring neural networks are a form of a cable of retelling components of neural network. In normal RNNs, this reiterating module have a humble arrangement is shown in Figure 5.4, such as a solo tanh layer.

LSTMs have a sequence-like construction for regular structure and a different structure in the repeating module as shown in Figure 5.5. So instead of single layer four layer is used in a different manner.

Every line represents a vector pointing from one node to another and the point-wise process is mentioned in pink circle and layers of neural networks with a yellow box.

More concentration is represented by merging lines and a content to be copied with a forking line and copied content moves to diverse positions. Figure 5.6 shows the various symbols used for various functions.

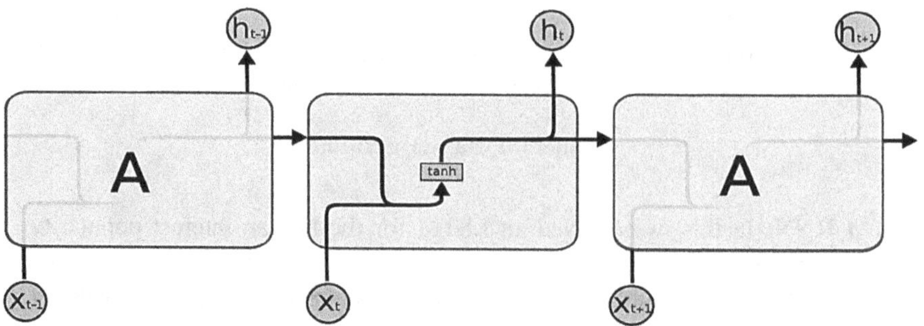

FIGURE 5.4 Single layer RNN in repeating module.

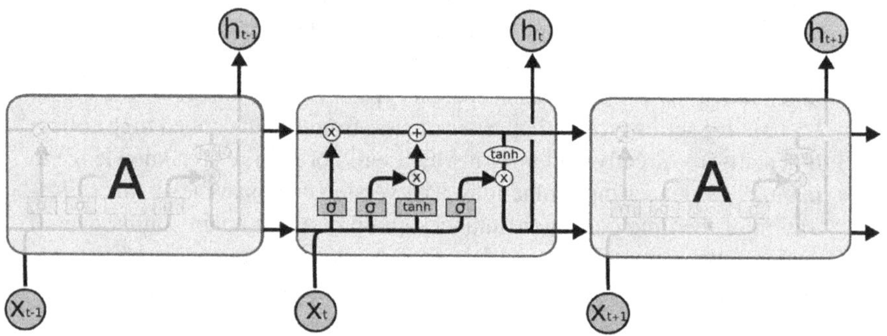

FIGURE 5.5 Four-layer LSTM in repeating module.

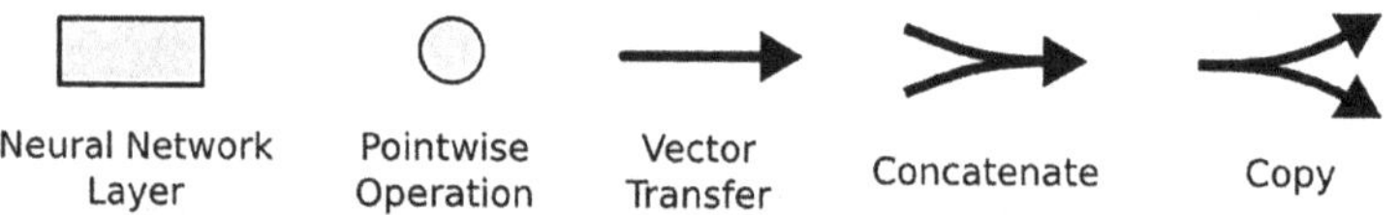

FIGURE 5.6 Various symbols used for various functions.

The primary thing on LSTM is a cell state representing like a conveyor belt is in the form of a horizontal line mentioned on top of the figure. This is connected in a straight manner toward the entire chain with little linear interactions. This is quite simple information that flows unchanged. The output of the sigmoid layer is between 0 and 1 that represents the amount of every component should enter into. 0 means nothing through and 1 means everything through. LSTM has three levels to secure and manage the cell state. The initial step in LSTM will identify the information to move away from one cell to another. The sigmoid layer is used for decision-making and is named as forget gate layer. The inputs are $h_{t-1}h_{t-1}$ and $x_t x_t$, and output is in between 00 and 11 on cell state $C_{t-1}C_{t-1}$. The output 11 shows to keep complete and a 00 denotes complete get rid of. The language model instance tries to expect additional words from previous ones as shown in Figure 5.7. In this, a cell state represents the present subject gender and to use a correct gender for the old subject.

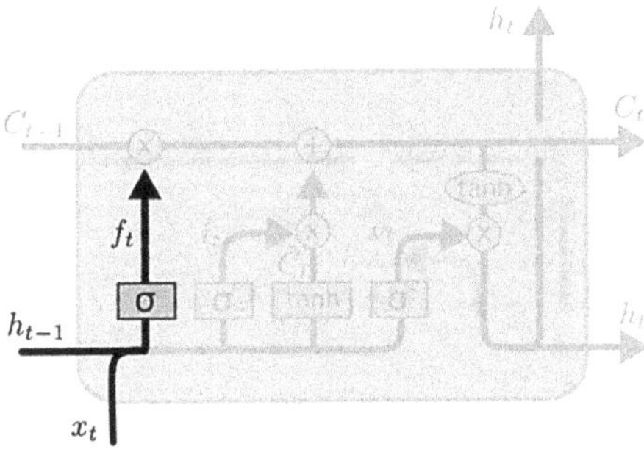

FIGURE 5.7 Language model instance.

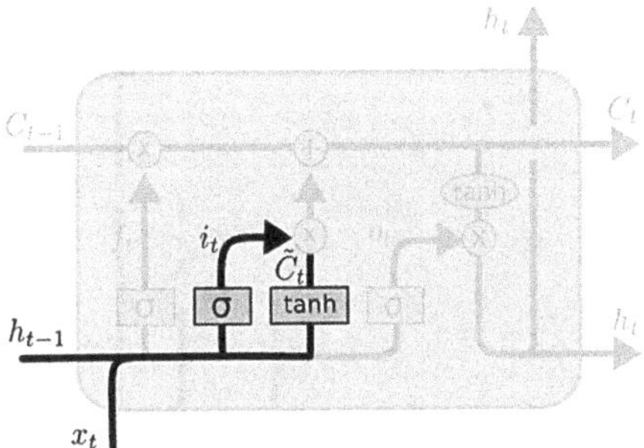

FIGURE 5.8 Steps in the layer updation.

The next stage is to identify the latest information that need to be stored into the cell. For this two parts are required. One is a sigmoid layer, in which the input gate layer identifies to be updated. Second is a tanh layer that has a new candidate values vector, C_t C_t, as added to state as shown in Figure 5.8. Next step is to combine the two states and update it by creating a new state. In the language model example, the new subject is added to the cell state by replacing the old one.

Now the old state, C_{t-1} C_{t-1}, is updated into C_t C_t the new state. The process that has to perform is decided in the earlier step and here it is multiplied by $f_t f_t$, and rejecting the operations earlier. Now add it∗C~tit∗C~t into it to get new value shown in Figure 5.9, which decides the amount to which the value needs to be scaled up. In the language model example, old subject information is dropped and the information of the new subject is added.

The output will be a filtered value from the cell state information. The executed sigmoid layer decides the value to be given at the output. The tanh pushes the values between –1–1 and 11, and will multiply the value to the output using the sigmoid gate which will be the desired value and is shown in Figure 5.10. In the example, the expected output may be a verb from a subject, that may be a singular or plural. Thus, the form of verb is conjugated and the output is obtained.

The explanation so far mentioned is a simple LSTM. All LSTMs are not like the simple one. It is like all the paper utilized in LSTM has a slight variant execution.

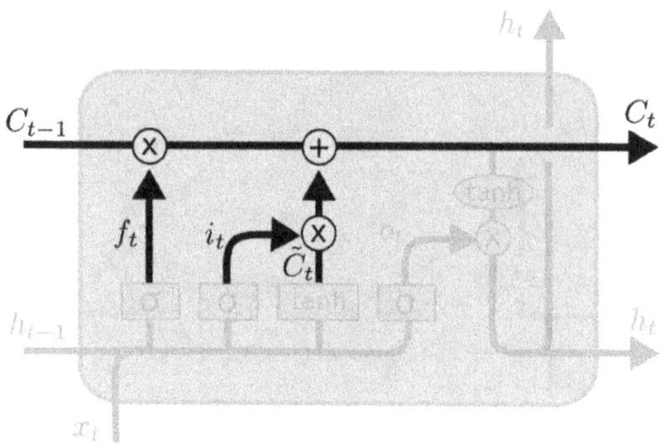

FIGURE 5.9 New value added in the operation due to updation.

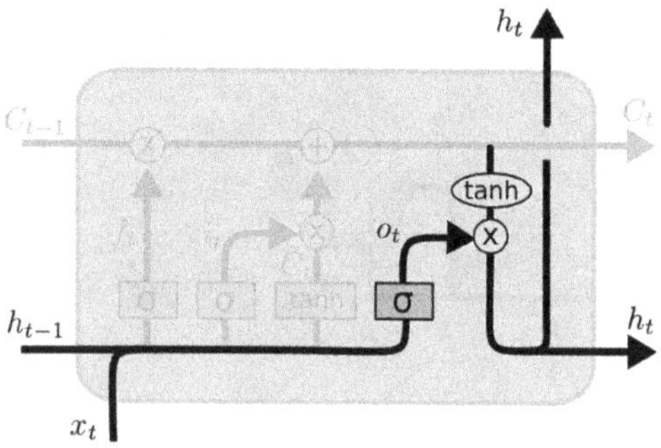

FIGURE 5.10 The diagrammatic representation of desirable output.

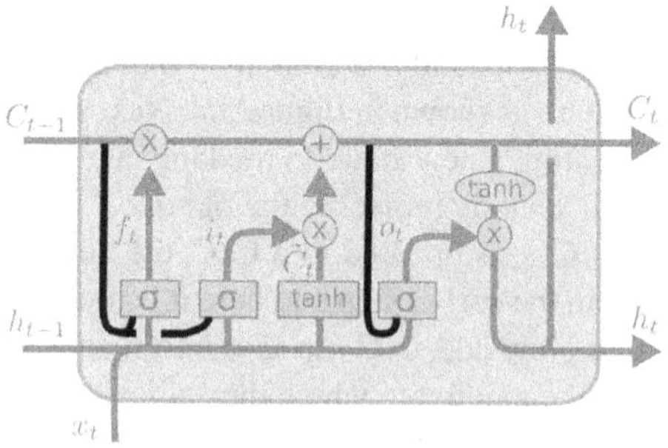

FIGURE 5.11 Diagrammatic representation of a gates peephole.

The value difference may be small but indicating them is important. Figure 5.11 shows a gates peephole, at the same time most of the publications will not show peepholes to others.

The usage of coupled forget and input gate is the other method. In order to identify what to forget and about the new information to be added, a combined approach is made. For this action, only few inputs are added as shown in Figure 5.12.

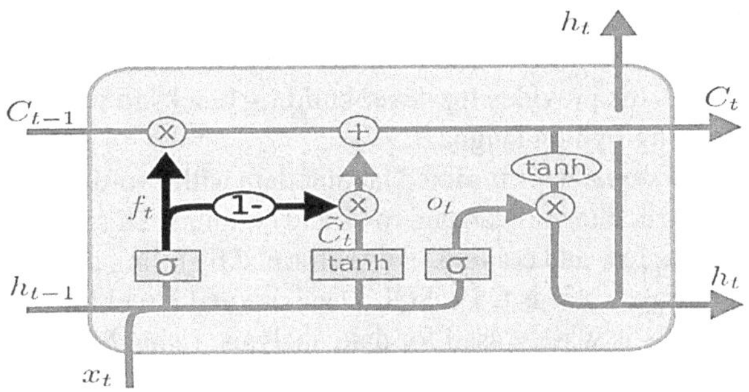

FIGURE 5.12 Schematic representation of combined approach.

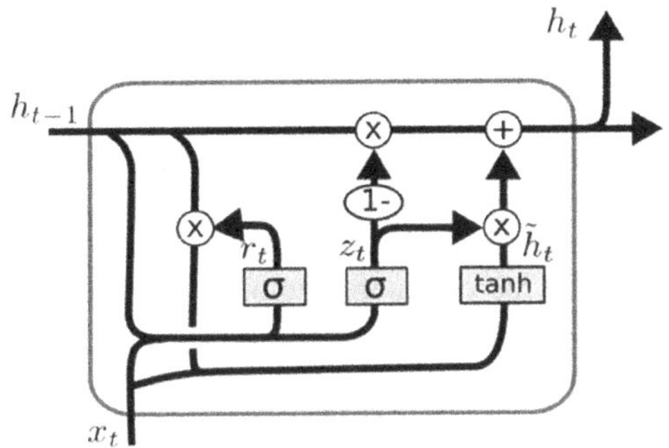

FIGURE 5.13 Gated Recurrent Unit (GRU).

Gated Recurrent Unit (GRU) is a different variance with slight modification in the LSTM. This combines the input gate and forget value into an update gate that merges the cell state, hidden state followed by few modifications. The output is easier than standard LSTM and is getting popular, which is shown in Figure 5.13.

Data Visualization: This is a graphical illustration of information's in the form of maps, charts, records, graphs, etc., to have better visualization and trends, outliers, and styles in records. In Big Data, this is important to react a large amount of information on decision-making. The bottom from expertise in various fields is used to obtain this.

Programming Language Used: Python is used as the language here, which is a high-level object-oriented interactive and interpreted language. It is extremely readable and uses keywords similar to English language.

Python libraries: The libraries used in our application are given below.

***NUMPY*:** This is a primary package to perform technical computation in Python language. This library gives various derived objects (such as masked arrays and matrices), multidimensional array object, and an assortment of routines for fast processes on arrays.

***PANDAS*:** This package gives flexible, fast, and sensitive data structures in order to work with—relational or—labeled data both informal and instinctive. This provides top-level building blocks to perform real data analysis using Python language.

***Pandas Data Frame*:** This is mixed tabular data with two-dimensional variable, data structure with labeled rows and columns. This has components such as data, row and column. This is formed by filling the datasets from present storage such as CSV, SQL Database and Excel file.

***Matplotlib*:** This is widely used for data analysis. Using Matplotlib, debatably the greatest graphing and data visualization can be done in an easy manner.

***Tensor Flow*:** This framework is a software library created by Google to develop ML and DL concepts in a simple way. This associates the mathematical evaluation for finding optimum solution in a simpler way.

5.5 DATA FLOW DIAGRAM

Data flow diagrams are portion of an organized archetypal and a graphical technique in the development of software that portrays information flow and transmutes that are realistic as a data transfer operation from input to output. The data flow figures are used for demonstrating the operator a graphical image which can be used for examination of the system.

These diagrams are separated into different levels. Level 0 is a software system, as shown in Figure 5.14 is a simple data flow diagram. Considering the number of inputs and outputs, the system will estimate the need for intermediate levels to obtain a better software model. Usually, a designer goes with three levels of data flow design structure.

Figure 5.14 illustrates a basic level 0 example of data flow diagram. In the diagram, the rectangle boxes represent external inputs or entities. These inputs may be a user code or a password. All the entities are correlated with the main system processes to get the necessary inputs for efficient processing. At level 0, no information about the process is needed, only input and output are required. Once the process is completed the information is given to the client or printer. Now the system is divided further into levels 1 and 2. These levels use same notation as in level 0 except some additional information is provided. The level 1 model is illustrated in Figure 5.15 and level 2 model is presented in Figure 5.16.

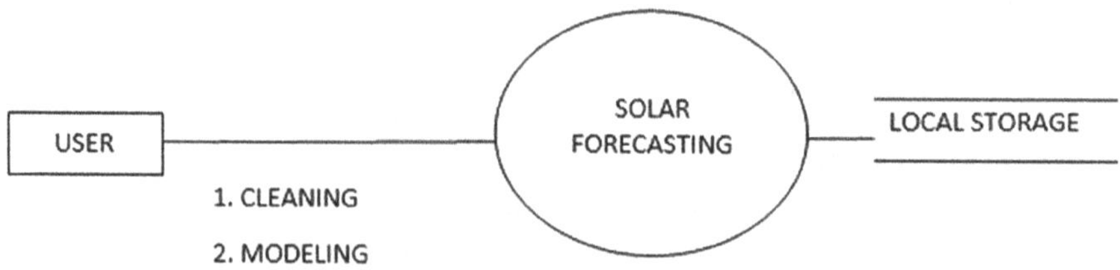

FIGURE 5.14 A simple illustration of a basic level 0.

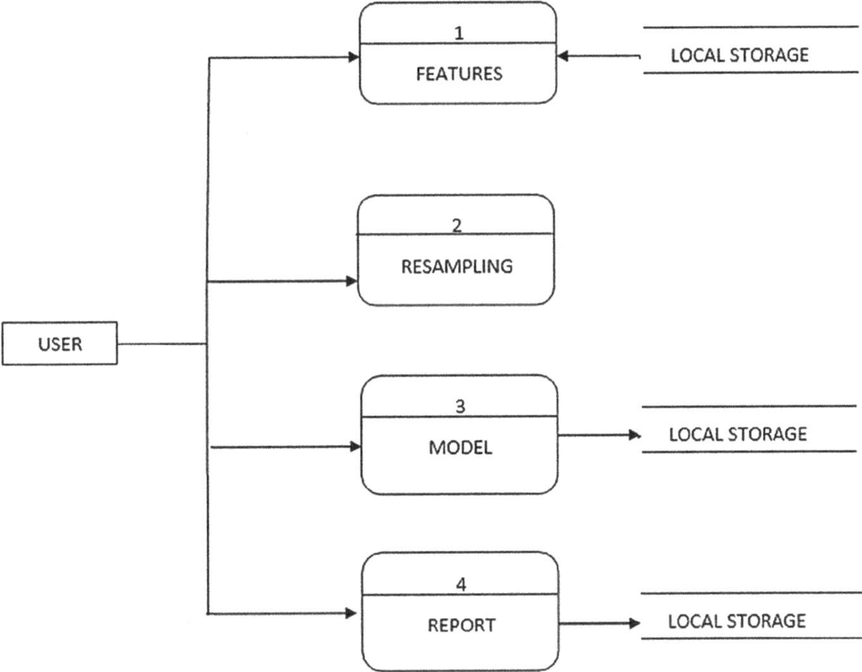

FIGURE 5.15 Level 1 model data flow diagram.

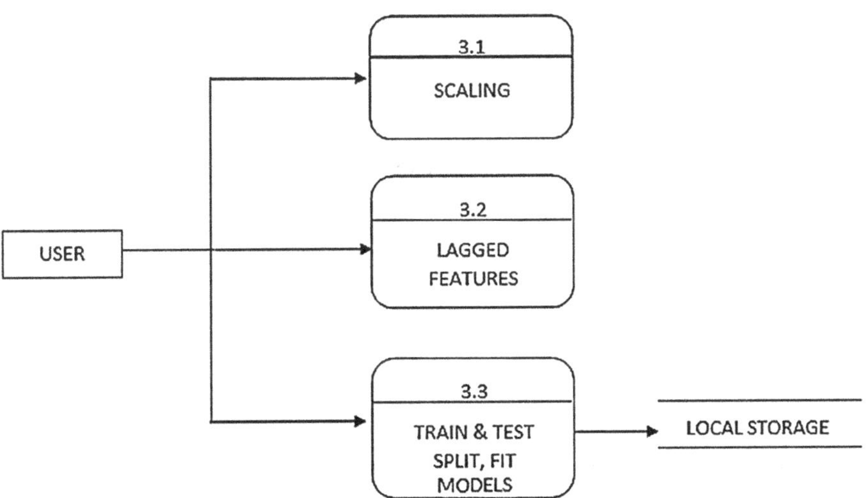

FIGURE 5.16 Level 2 model data flow diagram.

5.6 RESULTS AND DISCUSSION

5.6.1 Import Libraries and Read Data from CSE Files

NumPy arrays facilitate advanced mathematical and other types of operations on large numbers of data. Typically, such operations are executed more efficiently. Additionally, Python has the broader goal of becoming the most powerful and flexible open-source data analysis/manipulation tool available in any language. It is already well on its way toward this goal. Table 5.1 shows the library file imported from the CSE file which has the fields indicating the start period, end period, total period, air temperature, Azimuthal angle, cloud opacity, dew point temperature, DNI, DHI, GHI, etc.,

5.6.2 Retrieve the Year and Day of Year from Time Stamp Index

For the analysis, a large number of features are consolidated from other Python libraries like scikits.timeseries, as well as created a tremendous amount of new functionality for manipulating the time series data. Pandas DataFrame is a two-dimensional size-mutable, potentially heterogeneous tabular data structure with labeled axes mentioned in rows and columns. Table 5.2 lists and describes the year and day retrieved from the time stamp index with fields indicating capacity, dew point temperature DHI, DNI, perceptible water, relative humidity, snow depth, surface pressure, wind direction, wind speed, zenith angle, time stamp, etc.,

5.6.3 Create PST for Access Data over a Day

Pandas DataFrame will be created by loading the datasets from existing storage, such as SQL Database, CSV file and Excel file. Table 5.3 shows the PST creation for accessing the data over a day which has the fields such as surface pressure, wind direction, wind speed, zenith angle, time stamp, year, day of year, hours, minutes, etc.

5.6.4 Overall Performance of GHI, DHI, DNI

The graphical illustration provides better visualization and trends in the records. This is very much important and necessary for decision-making. Figure 5.17 shows the overall performance of GHI, DHI and DNI by time of day with respect to the received radiation range. From the graph, it is clear that the generated solar power radiation for DHI is highest during the noon time compared with the other two parameters.

5.7 CONCLUSION AND FUTURE SCOPE

Conclusion: Solar irradiance forecast has captured the eye of cutting-edge studies because of the requirement and hobby in renewable and inexperienced electricity. Accurate forecasting of sun irradiance is needed to apprehend the sun electricity angle of a region, thinking about the possibilities in addition to demanding situations associated with forecasting. The RNN fashions can effectively and precisely foresee as it should be expecting each day sun irradiance information. In this work, the RNN

TABLE 5.1
Import Libraries and Read Data from CSE Files

Period End	Period Start	Period	Air Temp	Albedo Daily	Azimuth	Cloud Capacity	Dew Point Temp	Dhi	Dni	Ghi	Gtifixed Tilt	Gti Tracking
2007-01-01T00 00 00Z	2007-01-01T00 30 00Z	PT3.0M	22.7	0.1	−113	0.0	18.1	2	2	2	2	2
2007-01-01T00 30 00Z	2007-01-01T00 0 00Z	PT3.0M	23.2	0.1	−114	7.0	18.4	17	10	17	18	16
2007-01-01T02 00 00Z	2007-01-01T00 30 00Z	PT3.0M	23.6	0.1	−116	37.5	18.7	59	5	60	60	62
2007-01-01T02 30 00Z	2007-01-01T02 0 00Z	PT3.0M	24	0.1	−119	18.6	19.0	136	88	166	176	220
2007-01-01T03 00 00Z	2007-01-01T02 30 00Z	PT3.0M	24.5	0.1	−122	1.9	19.2	139	430	304	358	567
2007-01-01T03 30 00Z	2007-01-01T03 00 00Z	PT3.0M	24.8	0.1	−125	0.0	19.3	161	548	419	487	684
2007-01-01T04 30 00Z	2007-01-01T03 30 00Z	PT3.0M	25.1	0.1	−130	0.0	19.2	181	614	522	600	743
2007-01-01T04 30 00Z	2007-01-01T04 00 00Z	PT3.0M	25.4	0.1	−135	0.0	19.0	196	665	615	701	782
2007-01-01T05 00 00Z	2007-01-01T04 30 00Z	PT3.0M	25.7	0.1	−142	0.0	18.9	206	704	694	786	782
2007-01-01T05 30 00Z	2007-01-01T05 00 00Z	PT3.0M	26.0	0.1	−150	0.0	18.8	215	730	757	852	782

TABLE 5.2
Retrieve the Year and Day of Year from Time Stamp Index

Cloud Capacity	Dew Point Temp	Dhi	Dni...	Precipitable Water	Relative Humidity	Snow Depth	Surface Pressure	Wind Direction 10m	Wind Speed 10m	Zenith	Time Stamp	Year	DOY
0.0	18.1	2	2	24.5	75.0	0.0	1012.9	40	2.7	94	2007-01-01 01 00 00-00 00	2007	1
7.0	18.4	17	10	24.6	74.4	0.0	1013.3	41	2.8	88	2007-01-01 01 30 00-00 00	2007	1
37.5	18.7	59	5	24.6	73.9	0.0	1013.7	42	2.9	81	2007-01-01 02 00 00-00 00	2007	1
18.6	19.0	136	88	24.7	73.3	0.0	1014.1	44	3.0	75	2007-01-01 02 30 00-00 00	2007	1
1.9	19.2	139	430	24.7	72.7	0.0	1014.5	45	3,1	68	2007-01-01 03 00 00-00 00	2007	1
0.0	19.3	161	548	24.8	71.5	0.0	1014.6	47	3.1	62	2007-01-01 03 30 00-00 00	2007	1

TABLE 5.3
Retrieve the Year and Day of Year from Time Stamp Index

Al Bedo Daily	Azimuth	Cloud Capacity	Dew Point Temp	Dhi	Dni...	Surface Pressure	Wind Direction 10m	Wind Speed 10m	Zenith	Time Stamp	Year	DOY	Hour	Min	PST
0.1	−113	0.0	18.1	2	2	1012.9	40	2.7	94	2007-01-01 01 00 00-00 00	2007	1	01	00	0100
0.1	−114	7.0	18.4	17	10	1013.3	41	2.8	88	2007-01-01 01 30 00-00 00	2007	1	01	30	0130
0.1	−116	37.5	18.7	59	5	1013.7	42	2.9	81	2007-01-01 02 00 00-00 00	2007	1	02	00	0200
0.1	−119	18.6	19.0	136	88	1014.1	44	3.0	75	2007-01-01 02 30 00-00 00	2007	1	02	30	0230
0.1	−122	1.9	19.2	139	430	1014.5	45	3.1	68	2007-01-01 03 00 00-00 00	2007	1	03	00	0300
0.1	−125	0.0	19.3	161	548	1014.6	47	3.1	62	2007-01-01 03 30 00-00 00	2007	1	03	30	0330

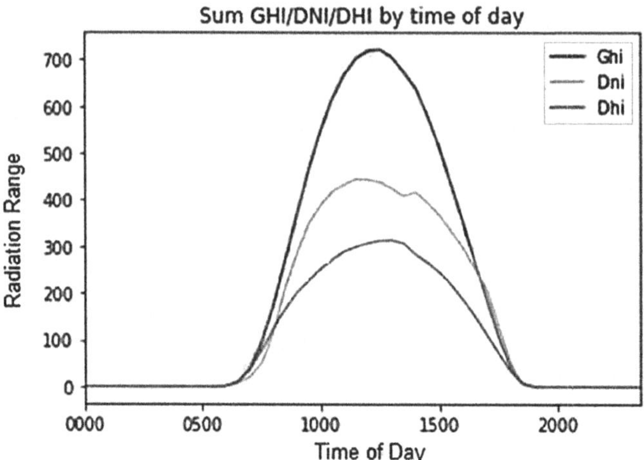

FIGURE 5.17 Overall performance of GHI, DHI, DNI.

fashions have been used to expect each day sun irradiance information with web page information of sun irradiance from Chennai place in India. A historic information set of sun irradiances during the last 10 years changed into used for education and checking out as it should be forecast sun irradiance on this study. For checking the validation and balance of the simulation outcomes, the goodness of suit of the version changed into examined the use of RMSE and coefficient of determination. The outcomes confirmed the functionality of the proposed method in supplying correct each day prediction of sun irradiance. The coefficient of determination (R2) changed into identical to 80% and 49% for each of the lagged features, respectively.

Future Scope: This assignment has included nearly all of the necessities. Further necessities and upgrades without problems can be carried out because the coding is specially based or modular in nature. Changing the present modules or including new modules can append upgrades. The destiny improvements recommended are: to expand this assignment at the fundamental of the requirement of the person; and currently handiest the minimum quantity of alternatives is being furnished to person and it predicted us to growth many alternatives to the person at large.

REFERENCES

1. https://www.tangedco.gov.in/chapter2.html.
2. S. Sobri, S. Koohi-Kamali, and N. A. Rahim, "Solar Photovoltaic Generation Forecasting Methods: A Review", *Energy Conversion and Management*, 156, 459–497, 2018.
3. Ö. Ayvazoğluyüksel, and Ü. B. Filik, "Estimation Methods of Global Solar Radiation, Cell Temperature and Solar Power Forecasting: A Review And Case Study in Eskişehir", *Renewable and Sustainable Energy Reviews*, 91, 639–653, 2018.
4. E. H. Abdelhakim, and B. Abdennaser, "Solar Photovoltaic Power Forecasting", *Journal of Electrical and Computer Engineering*, 2020, 21, 2020. https://doi.org/10.1155/2020/8819925

5. D. O'Leary, and J. Kubby, "Feature Selection and ANN Solar Power Prediction", *Journal of Renewable Energy*, 2017, 7, 2017. https://doi.org/10.1155/2017/2437387
6. J. Jose Anand R. P. Perinbam, and D. Meganathan, "Design of GA-Based Routing in Biomedical Wireless Sensor Networks", *International Journal of Applied Engineering Research*, 10(4), 9281–9292, 2015.
7. P. Prem Kumar, K. Duraiswamy, and J. Anand, "An Optimized Device Sizing of Analog Circuits Using Genetic Algorithm" *European Journal of Scientific Research*, 69, 3, 441–448, 2012,
8. Z. Zhen, W. Zheng, F. Wang, Z. Mi, and K. Li, "Research on a Cloud Image Forecasting Approach for Solar Power Forecasting, *Energy Procedia*, 142, 362–368, 2017.
9. B. Sivaneasan, C. Y. Yu, and K. P. Goh, "Solar Forecasting Using ANN with Fuzzy Logic Pre-Processing", *Energy Procedia*, 143, 727–832, 19–21 July 2017.
10. A. Roy, A. Ramanan, B. Kumar, N. Kumaar, C. A. Abraham, et al., "Day Ahead Solar PV Power Forecasting Based on a Combination of Statistical and Physical Modelling Utilizing NWP Data for Solar Parks in India", *2nd International Conference on Large Scale Grid Integration of Renewable Energy in India*, New Delhi, India, 4–6 Sep 2019.
11. J. Anand, A. Jones, T. K. Sandhya, and K. Besna, "Preserving National Animal Using Wireless Sensor Network Based Hotspot Algorithm", *Proceedings of 2013 IEEE International Conference on Green High Performance Computing*, 1–6, 14–15 March 2013.
12. M. Roulston, D. Kaplan, J. Hardenberg, and L. Smith, "Using Medium range Weather Forecasts to Improve the Value of Wind Energy Production", *Renewable Energy*, 28(4), 585–602, 2003.
13. A. Gensler, J. Henze, B. Sick, and N. Raabe, "Deep Learning for Solar Power Forecasting: An Approach Using Auto Encoder and LSTM Neural Networks", In *2016 IEEE International Conference on Systems, Man, and Cybernetics*, 2858–2865, Oct 2016.
14. S. Makrikakis, E. Spiliotis, and V. Assimakopoulos, "Statistical and Machine Learning Forecasting Methods: Concerns and Ways Forward", *PLOS ONE*, 13(3), 1–26, March 2018.
15. R. Perez, S. Kivalov, J. Schlemmer, K. Hemker, D. Renne, and T. E. Hoff, "Validation of Short and Medium Term Operational Solar Radiations Forecasts in the US", *Solar Energy*, 84, 2161–2172, 2010.
16. J. Anand, and J. Raja Paul Perinbam, "Automatic Irrigation System using Fuzzy Logic", *AE International Journal of Multidisciplinary Research*, 2(8), 1–9, August 2014.
17. A. Dairi, F. Harrou, Y. Sun, and S. Khadraoui, "Short-Term Forecasting of Photovoltaic Solar Power Production Using Variational Auto-Encoder Driven Deep Learning Approach", *Applied Sciences*, 10, 8400, 2020.
18. Zhang G., and Guo J., "A Novel Method for Hourly Electricity Demand Forecasting", *IEEE Transactions on Power System*, 33, 1351–1363, 2020.
19. M. Rana, and A. Rahman, "Multiple Steps Ahead Solar Photovoltaic Power Forecasting Based on Univariate Machine Learning Models and Data Re-Sampling", *Sustainable Energy Grids and Networks*, 21, 100286, 2020.
20. M. Geethalakshmi, J. A. Kanimozhiraman, R. Partheepan, and S. Santhosh, "Optimal Routing Path using Trident Form in Wearable Biomedical Wireless Sensor Networks", *Turkish Online Journal of Qualitative Inquiry*, 12(7), 5134–5143, 2021.
21. X. Zhang, Y. Li, S. Lu, H. F. Hamann, B. Hodge, and B. Lehman, "A Solar Time Based Analog Ensemble Method for Regional Solar Power Forecasting", *IEEE Transaction on Sustainable Energy*, 10, 268–279, 2019.

22. M. J. Sanjari, H. B. Goori, and N. C. Nair, "Power Generation Forecast of Hybrid PV – Wind System", *IEEE Transactions on Sustainable Energy*, 11, 703–712, 2020.
23. J. R. Andrade, and R. J. Bessa, "Improving Renewable Energy Forecasting with a Grid of Numerical Weather Predictions", *IEEE Transactions on Sustainable Energy*, 8, 1571–1580, 2017.
24. M. V. Arokiamary, and J. Anand, "Analysis of Dynamic Interference Constraints in Cognitive Radio Cloud Networks", *International Journal of Advanced Research in Science, Communication and Technology*, 6(1), 815–823, June 2021.
25. P. A. G. M. Amarasinghe, N. S. Abeygunawardana, T. N. Jayasekara, E. A. J. P. Edirisinghe, and S. K. Abeygunawardane, "Ensemble Models for Solar Power Forecasting – A Weather Classification Approach", *AIMS Energy*, 8(2), 252–271, 2020.
26. S.-G. Kim, J.-Y. Jung, and M. K. Sim, "A Two-Step Approach to Solar Power Generation Prediction Based on Weather Data Using Machine Learning", *Sustainability*, 11, 1501, 2019.

6 Aerodynamics of Wind Turbine

Ayushi Rawat
Trane Technologies, Bengaluru, India

CONTENTS

6.1 INTRODUCTION

Energy can neither be created nor be destroyed; it can only be converted from one form to another. Energy consumption is increasing on the daily basis, although there has been slowdown in energy consumption in 2019 (+0.8%) as a result of slower economic growth. In 2019, world energy production continued growing (+1.5%); however for year 2020, it went down by 3.5% [1]. To facilitate energy demand without compromising the needs of future generation, we should slowly rely more on sustainable energies.

However, in present times global energy demand is met from fossil fuels (61.3%) and 35.2% as nuclear power and renewable energy which comprises hydro, solar and wind as shown in Figure 6.1 [1]. Studies on renewable energy would help surge in renewable energy figures. This chapter focuses on one such energy, wind energy, which is clean and massively moves in atmosphere.

DOI: 10.1201/b23013-6

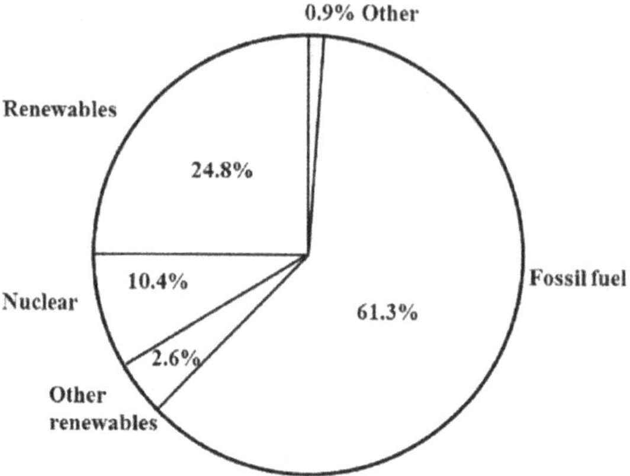

FIGURE 6.1 Outlook of world electricity generation.

Through this chapter, we will understand the energy available in the wind and how the energy is converted to generate electricity. There is always a limitation of extracting energy that is called as Betz limit, and to obtain energy various design parameters play an important role. There are three different aerofoil theories which are basically a mathematical concept to understand the lift and drag force generation in the blade and their relationship with different parameters.

6.2 HISTORY

People have been using wind energy for thousands of years from now. In 5000 BC people used it to propel boats; later humans harnessed wind for various other purposes, wind-powered water pumps, and grain grinding mills. In 1887, first known wind turbine was built in Scotland to produce electricity by Professor James Blyth, which was installed in his cottage and was used to charge accumulator which powered the cottage, and thus became the first house to have electricity supplied by wind energy. Wind energy use expanded when there was a shortage of oil in 1970s around the world and also due to environmental concerns. This shortage created an interest in developing ways to alternative energy sources. Energy harnessed by wind is emission free, as it does not pollute environment (EIA, 2021). An individual wind turbine has a small impact; this is why a large number of wind turbines are installed in an open land or on ocean.

6.3 CLASSIFICATION OF WIND TURBINES

Based on the energy conversion, the wind turbines are classified as follows:

1- Windmill: When the kinetic energy is converted to mechanical energy, it is directly used by the machines such as pumps, grinding machines, etc. This type of wind turbine is called windmill.
2- Wind turbines: When the kinetic energy is converted to electrical energy, the machine is called wind turbine.

Wind turbines are classified into two categories:

1- Horizontal-axis wind turbine (HAWT)
2- Vertical-axis wind turbine (VAWT)

6.3.1 Horizontal-Axis Wind Turbine

The horizontal-axis wind turbine has an axis parallel to the ground. They are mostly used due to their strength and efficiency. The generator and gearbox are placed on the top of the tower; this is because the base of the tower has to be strong. There are single-bladed, two-bladed and multi-bladed wind turbines; most popular are multi-bladed wind turbines (three-bladed turbines). They can further be classified based on the direction of receiving air, i.e., upwind and downwind turbine [2].

Upwind has its rotors facing the wind. In this turbine, rotor needs to be fixed with some distance. In addition to this, the yaw mechanism is required to keep the rotor facing the wind. However, downwind has rotor facing the lee-side of the tower [3]. It does not require the yaw mechanism and the basic advantage of upwind turbine is reduced weight and good dynamic stability as shown in Figure 6.2.

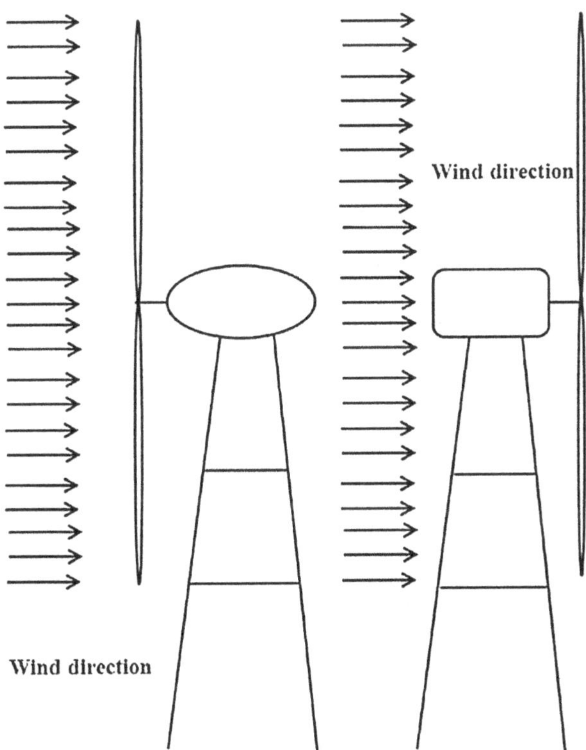

FIGURE 6.2 Schematic diagram of upwind and downwind turbine.

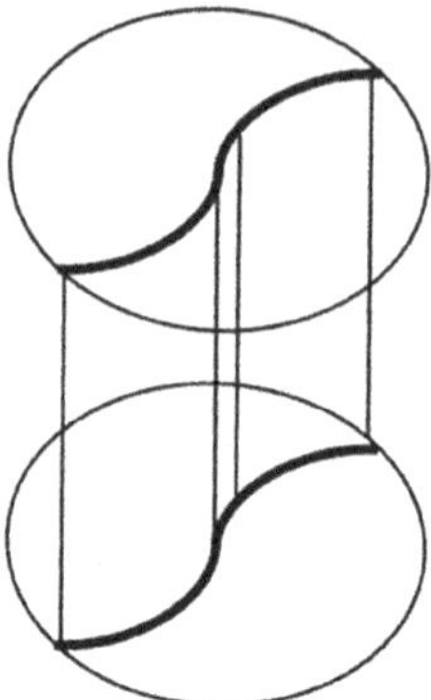

FIGURE 6.3 Savonius vertical-axis wind turbine.

6.3.2 Vertical-Axis Wind Turbine

It has its blade rotating on an axis perpendicular to the ground. In this turbine, wind can pass through from any direction and the gearbox and generator are placed on the ground. In this way, yaw mechanism can be eliminated [4]. One of its advantages is that its maintenance is easier since it is mounted on the ground (gearbox and generator). This type of turbine experiences drags which results in lower efficiency and also they require some external mechanism (push mechanism) to start it once stopped.

There are various types of VAWT:

1- Savonius turbine: This VAWT when viewed from top is of S-shaped.
   It is used for pumping water, grinding grains, and various other tasks. Drag force is the key element for this turbine; concave surfaces have more drag than convex as shown in Figure 6.3. It rotates at a slow speed and therefore is not used for generating electricity.
2- Darrieus turbine: This is one of the most common VAWT, named after French engineer George Darrieus.

It has C-shaped rotor with two or three blades that also appear as an egg beater as depicted in Figure 6.4. It works by the lift generated from aerofoils, arrangements of blades and minimises the bending stress on it. Giromill wind turbine is one of the variations of Darrieus wind turbine. It has a set of straight vertical blades attached to the vertical axis.

6.4 POWER AVAILABLE IN WIND

Wind is a clean and green energy system; therefore, it becomes one of the vital forms of sustainable energy resource. Energy available in wind is kinetic which is transformed to mechanical to generate electrical energy with the help of wind turbine. The efficiency of converting wind energy to useful energy forms depends on the rotor's interaction wind stream [5].

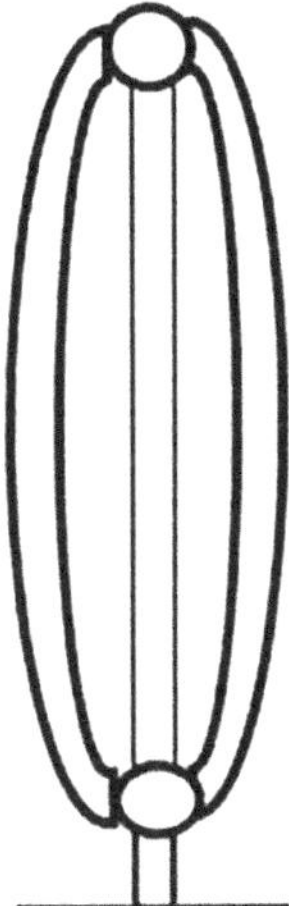

FIGURE 6.4 Darrieus wind turbine.

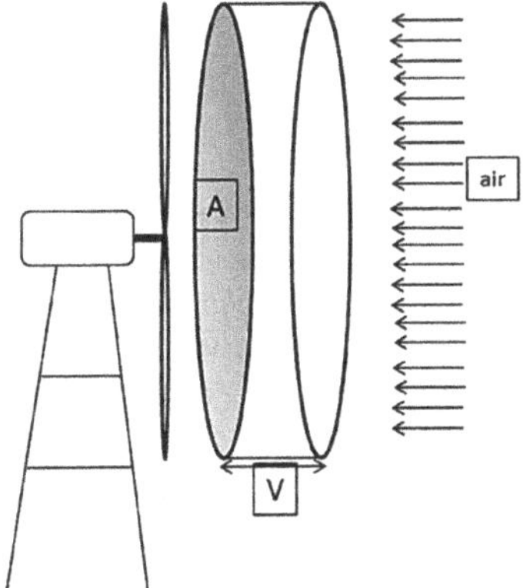

FIGURE 6.5 Representation of energy developed from wind.

Power available in wind (or air) of mass m, moving with velocity V, is given by

$$E = \frac{1}{2} mV^2 \tag{6.1}$$

The wind energy can be calculated by considering a cylindrical profile having thickness same as wind velocity as shown in Equation 6.2, as shown in Figure 6.5.

$$E = \frac{1}{2} \rho A V^3 \tag{6.2}$$

Since $\rho = \frac{m}{\upsilon}$ and $\upsilon = A^*t$, where ρ = density of air

A = swept area
υ = volume of air
t = thickness = V (wind velocity)

So from Equation 6.2, wind energy depends on:

1- Velocity of air: The amount of energy in wind varies as cubic, so a slight change in velocity will result in prominent energy.
2- Swept Area: Larger the area covered by blades, more the energy. Length of the blade is directly proportional to the area. $A = \pi r^2$, where r is the radius of rotor.
3- Density of air: Higher the density more is the energy produced. It varies with elevation and temperature. Turbine produces more energy when it is located at lower elevation with colder temperature range. Based on temperature (T) and elevation (z), density of air can also be calculated as

$$\rho = \frac{353.049}{T} e^{\left(-0.034\frac{z}{T}\right)} \tag{6.3}$$

6.5 WIND TURBINE POWER AND ITS TORQUE

The amount of power available in wind scales cannot be 100% extracted. Some part of it passes through the blades which lead in production. Actual power produced depends on the efficiency of the rotor to convert wind energy. This efficiency is called power coefficient C_p, ratio of actual power developed by rotor to theoretical power, P_T, available in wind as shown in the following equation:

$$C_p = \frac{2P_T}{\rho A V^3} \tag{6.4}$$

There are many factors on which power coefficient depends on such as turbine blades, wind speed, turbine blade angle, rotation of turbine, etc.

For wind turbine to be 100% efficient, the rotor would have to be disk shape which in turn will not rotate, and thus no generation of electricity. In the wind energy conversion system, we come across various losses that define the efficiency of the component to convert on form of energy to desired output. In Figure 6.6, we see that power coefficient is the product of turbine, mechanical and electrical efficiency [6].

Albert Betz was a German physicist who concluded that no wind turbine can convert more than 59.3% of wind energy to mechanical. This is known as the Betz limit and theoretically it is the maximum limit of C_p [7].

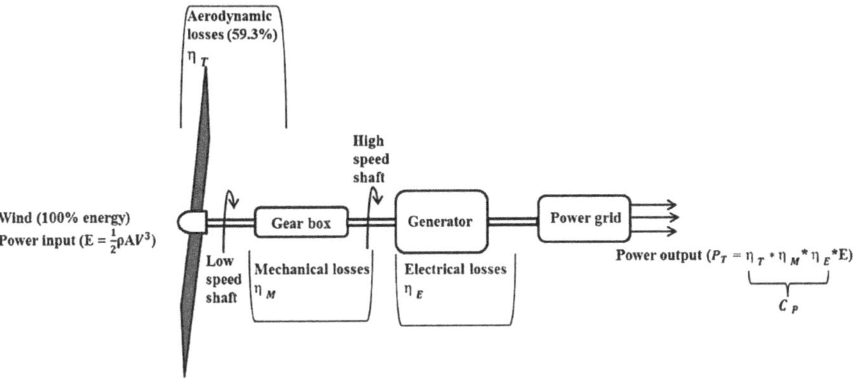

FIGURE 6.6 Power flow in wind turbine.

Example of Betz Limit

In the above figure, it is shown that only 70% of Betz limit is converted by turbine to useful electrical energy. Therefore, this turbine will convert only 41% of available wind energy to electrical energy. $0.7 \times 0.593 = 0.415$ Good turbine are those which ranges from 35%–45%

When the blades rotate, they experience a drag force, F_D, as shown in the following equation:

$$FD = \frac{1}{2}\rho AV^2 \tag{6.5}$$

So, torque of wind turbine with radius r is

$$T = F_D * r \tag{6.6}$$

The above torque is the maximum theoretical torque, but only fraction of it is developed. This is termed torque coefficient C_T. It is the ratio of actual torque T_T developed to the theoretical torque as shown in the following equation:

$$C_T = \frac{2T_T}{\rho AV^2 r} \tag{6.7}$$

Tip speed (Equation 6.8) ratio is another extremely important parameter which depicts about the performance of turbine [8]. It is the ratio of velocity of rotor tip to the wind velocity:

$$TSR(\lambda) = \frac{\text{tip speed of rotor}}{\text{wind velocity}} \tag{6.8}$$

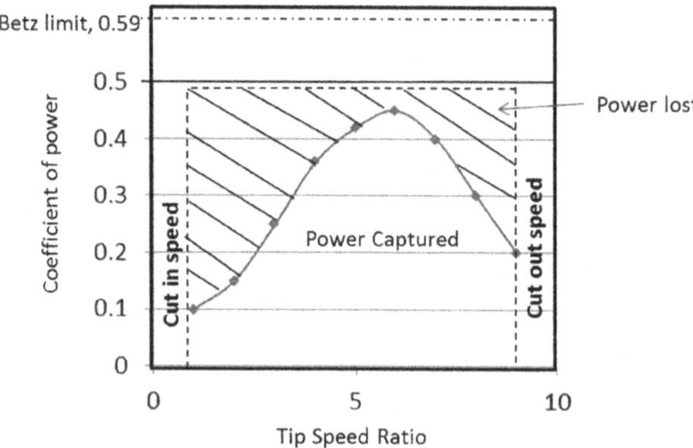

FIGURE 6.7 Relation between power coefficient and tip speed ratio.

Blades spin fast at the tip than at centre. Let us understand through an example the importance of tip speed ratio. If the rotor is rotating slowly than most of the air passes through the gap between the blades, hence no conversion of energy. Similarly if rotor spins fast, blades will rotate at a higher pace. As the blades rotate, it creates turbulence and if the next blade arrives sooner it will hit the turbulent air thus deflecting the wind stream which in turns causes loss of energy [9].

In both the above cases result in poor power coefficient, it is better to design wind turbine with optimum tip speed ratio as shown in Equation 6.9.

$$\lambda = \frac{r\omega}{V} = \frac{2\pi fr}{V} \tag{6.9}$$

where ω is angular speed and f is frequency of rotor.

As shown in Figure 6.7, it is very important to design the turbine at an optimum value of tip speed ratio with respect to the power coefficient.

Cut in speed: It is the speed in which turbines are designed to start function. It usually ranges from 3 to 5 m/s.

Cut out speed: It is the speed at which turbines stop functioning, which is above 25m/s in order to avoid any damage to wind turbine at a higher speed [10].

Example: The turbine rated at 1.75 MW rated power at 14 m/s rated wind speed. Rotor diameter is 65 m and rotational speed ranges from12.9 to 21.2 rpm. Calculate the tip speed ratio range.

Solution:

Given, d = 65 m, V = 14m/s, N_1 =12.9 rpm and N_2 = 21.2 rpm

$$\lambda = \frac{2\pi fr}{V}$$

$$\lambda 1 = \frac{2\pi f_1 r}{V} = \frac{2\pi * 12.9 * 65}{14 * 60 * 2} = 3.13$$

$$\lambda 2 = \frac{2\pi f_2 r}{V} = \frac{2\pi * 21.2 * 65}{14 * 60 * 2} = 5.15$$

$$\lambda \approx 3 \text{ to } 5$$

POINTS TO REMEMBER: If TSR is 1 or above it means blades experience a lift which helps blades to spin fast and if TSR is less than 1 then there is lot of drag.

6.6 AERODYNAMICS OF WIND TURBINE

If any object is submerged in a fluid (liquid or air), it experiences two forces, i.e., drag and lift. The blades are designed as aerofoil which helps in extracting more energy from airstream. Aerofoil is used in wind turbines and is from NACA series (National Advisory Committee for Aeronautics). NACA series has different specifications defined by numbers [11].

NACA 4 digit specification

NACA ZXCV example NACA 2415

Where Z is the maximum camber in the chord line (divided by 100), here maximum camber is 0.02 or 2% of chord,
X is the location of camber from the leading edge (divided by 10), here 0.4 or 40% of chord and
CV is the thickness (divided by 100), here 0.15 or 15% of chord.

NACA 5 digit specification

NACA VWXYZ example NACA 23012

Where V is the lift coefficient multiplied by 1.5 times (divided by 10), here C_L is 0.3
WX is the location of maximum camber from leading edge multiplied by 0.5 (divided by 100), here 0.15 or 15% and
YZ is thickness (divided by 100), here 0.12 or 12% of chord.

Aerofoil is the cross-section of wind turbine. The mean camber line is the locus of points, equidistant from upper and lower surfaces. Leading and trailing edge are the front point of the blade is leading edge and the point at the back of the blade is the trailing edge. Chord Line is the straight line connecting the leading and trailing edge. Chord Length is the distance between leading and trailing edge along the chord line. Angle of attack is the angle between chord line and the direction of relative wind [12].

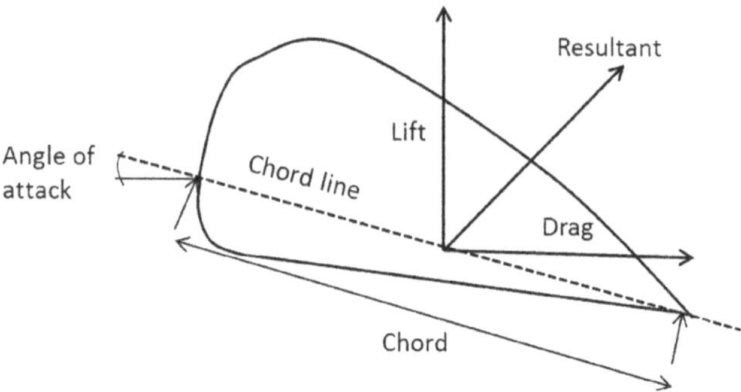

FIGURE 6.8 Aerofoil parameters.

When any object or aerofoil is submerged in the fluid (liquid or gaseous), here wind stream it experiences a lift and drag force. The air when hits the aerofoil results in lift force perpendicular to the wind velocity and parallel to it is drag force as shown in Figure 6.8. The upper surface of aerofoil cross-section experience higher velocity, this is because of circulation is created on the upper surface and consequently the lower surface has lower velocity. Assuming we have irrotational flow we apply Bernoulli's theorem in order to have equality we have lower pressure on the upper surface and higher surface on the lower surface. This pressure difference results in force which has a component lift and drag force. Forces are given as follows:

$$\text{Lift force } F_L = C_L \frac{1}{2} \rho A V^2 \tag{6.9}$$

$$\text{Drag force } F_D = C_D \frac{1}{2} \rho A V^2 \tag{6.10}$$

Here A is the platform area (B*C), as depicted in Figure 6.9. C_L and C_D are coefficient of lift and drag. When angle of attack increases lift force increases until it reaches maximum value and then it rapidly start decreasing with that drag force

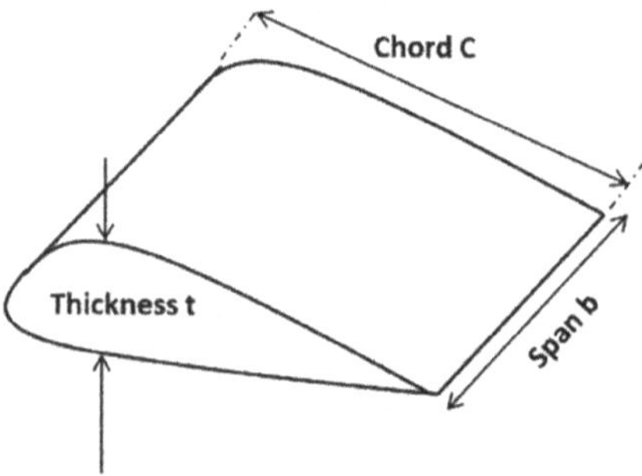

FIGURE 6.9 Representation of platform area.

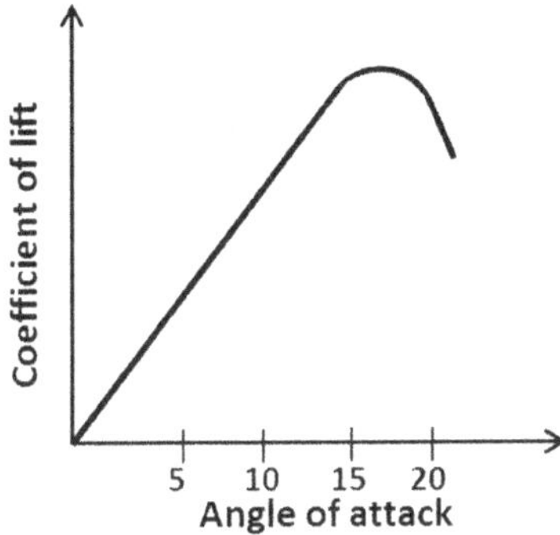

FIGURE 6.10 Relation between coefficients of lift with angle of attack.

increase. This is because once the aerofoil reaches its maximum angle of attack, the boundary layer gets separated known as stall. Hence it is important to have aerofoil at optimum angle of attack to extract more energy. Another factor on which coefficient of lift or drag depends is Reynolds number as shown in Equation 6.11 (based on the chord length):

$$Re = \frac{\rho VC}{\mu} \tag{6.11}$$

where ρ is the density of fluid, V is the velocity, C is the chord length and μ is the dynamic viscosity.

The lift coefficient increases linearly with angle of attack reaches maximum at about 16° to 17° and then starts decreasing drastically as depicted in Figure 6.10.

At zero angle of attack:

1- Lift coefficient is zero for symmetrical aerofoils
2- nonzero for no symmetrical ones with greater curvature at top.

6.6.1 Aerofoil Theory

The velocity at which wind stream approaches the blades may differ based on different blade sections. In order to maintain optimum angle of attack throughout all blade sections, we have aerofoil theories that will help in designing the blade [13]. Following are the theories:

1- Axial Momentum Theory
2- Blade Element Theory
3- Strip Theory

6.6.1.1 Axial Momentum Theory

Assume a wind turbine blade with area A and similarly consider sections 1–1 and 2–2 having areas A_1 & A_2 with velocities V_1 & V_2, respectively, as shown in Figure 6.11.

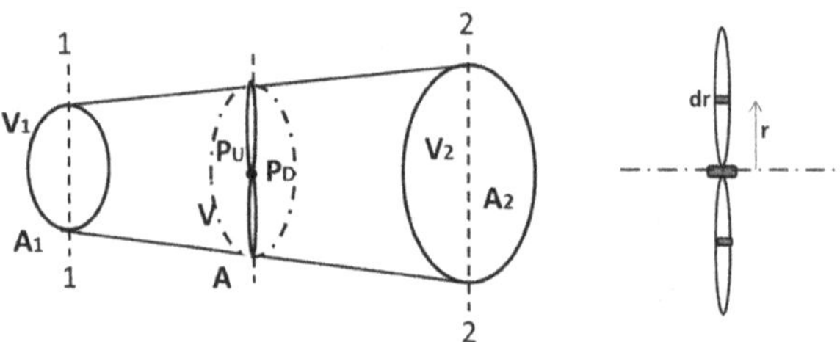

FIGURE 6.11 Representation of stream tube around wind turbine.

Assumptions:

1- Flow is incompressible and homogeneous.
2- Ideal flow conditions.
3- Static pressure.

Applying law of conservation of mass depicted in Equation 6.12 and 13:

$$\rho A_1 V_1 = \rho A V = \rho A_2 V_2 \tag{6.12}$$

$$A_1 V_1 = A V = A_2 V_2 \tag{6.13}$$

Thrust force experienced by rotor is also given as difference in momentum:

$$F = \rho A_1 V_1^{\,2} - \rho A_2 V_2^{\,2} \tag{6.14}$$

$$F = \rho A V (V_1 - V_2) \tag{6.15}$$

[from eqn. 6.12]

Thrust force can also be given as

$$F = (P_U - P_D) A$$

Applying Bernoulli's Theorem to sections 1–1 & 2–2 and considering static atmospheric pressure P at both the section:

$$P + \frac{\rho V_1^2}{2} = P_U + + \frac{\rho V^2}{2} \tag{6.16}$$

$$P + \frac{\rho V_2^2}{2} = P_D + + \frac{\rho V^2}{2} \tag{6.17}$$

From above eqn. we get Equation 6.18:

$$P_U - P_D = \frac{\rho}{2}\left(V_1^2 - V_2^2\right) \tag{6.18}$$

Equating this to thrust force,

$$F = \frac{\rho A}{2}\left(V_1^2 - V_2^2\right) \tag{6.19}$$

While comparing Equations 6.19 and 6.15, we get Equation 6.20:

$$V = \frac{V_1 + V_2}{2} \tag{6.20}$$

The wind velocity at upstream decreases by some factor called as axial induction factor a,

$$a = \frac{V_1 - V}{V_1}$$

$$V = (1-a)V_1 \tag{6.21}$$

$$V_2 = V_1(1-2a)$$

Power developed by turbine is given as shown in Equation 6.22:

$$P_T = \frac{1}{2}\rho AV\left(V_1^2 - V_2^2\right) \tag{6.22}$$

Substituting value of V and V_2 in the above equation we get Equation 6.23:

$$P_T = \frac{1}{2}\rho AV_1^3\, 4a(1-a)^2 \tag{6.23}$$

where

coefficient of power is as shown in Equation 6.24:

$$C_P = \frac{P_T}{P_{avail}} = 4a(1-a)^2 \tag{6.24}$$

Since $P_{avail} = \frac{1 \rho A V_1^3}{2}$

Similarly for coefficient of thrust, $C_T = \frac{F_T}{F_{avail}}$ we get Equation 6.25:

$$F_T = \frac{\rho A}{2}\left(V_1^2 - V_2^2\right) = 2a\rho\, A V_1^2 (1-a)$$

$$F_{avail} = \frac{1 \rho A V_1^2}{2}$$

$$C_T = 4a(1-a) \tag{6.25}$$

For maximum power developed $\frac{dC_P}{da} = 0$, we get Equation 6.26:

$$a = 1/3$$

$$P_{Tmax} = \frac{1}{2}\rho A V_1^3 \frac{16}{27} \tag{6.26}$$

16/27 is the limit of power coefficient which is known as Betz limit.

Considering the tangential flow behind the rotor, tangential induction factor $a_1$is given as Equation 6.27:

$$a_1 = \frac{U_\theta}{2r\omega} \tag{6.27}$$

U_Θ = induced tangential angular velocity of flow
ω = angular rotation of rotor.

Assuming annular tube of thickness dr at r distance from root of the blade, area is given as

$$A = 2\pi r\, dr$$

Force experienced by annular strip is given as Equation 6.28:

$$dF = \frac{dm\,(V_1 - V_2)}{dt} = \rho\, 2\pi r\, V\left(V_1 - V_2\right)dr \tag{6.28}$$

Equating V and V_2 in Equation 6.28, we get Equation 6.29 as shown below:

$$dF = 4a(1-a)\rho\, \pi r\, V_1^2\, dr \tag{6.29}$$

Torque on the annular element:

$$dT = 2\pi r^2 \rho U_\theta \, dr \tag{6.30}$$

Equating 6.21 and 6.27 in 6.30, we get Equation 6.31:

$$dT = 4\pi r^3 a_1 \left(1 - a\right) \rho V_1 \omega \, dr \tag{6.31}$$

Power developed by rotor is given by integrating the product of torque and angular velocity of flow as shown in equation 6.32:

$$P = \int_0^R r\omega 4\pi r3a1\left(1 - a\right) \rho V1\omega \, dr \tag{6.32}$$

$$C_P = \frac{2P}{\rho A V^3}$$

Example: A 65 m diameter rotor experiences an undisturbed wind speed of 15 m/s. If it is operating at maximum C_p and air density is 1.165 kg/m³. Calculate the following:

(1) Velocity at the turbine at section, (2) Velocity at the downstream, (3) Pressure difference across the disc using actuator disc analysis, (4) Thrust on the rotor at this condition, (5) Maximum power developed by the rotor, (6) Coefficient of thrust, (7) Blade tip speed ratio, (8) Maximum angular velocity and (9) Torque developed at the maximum power.

Solution:

Given: D = 65 m, V_1 = 15 m/s, C_P = 16/27 (operating at max), so $a = \frac{V_1 - V}{V_1}$

1- At maximum operating condition a = 1/3

$$1/3 = \frac{15 - V}{15}$$

$$V = 10 m/s$$

2- $V_2 = 15\left(1 - \frac{2}{3}\right)$

$$V_2 = 5 m/s$$

$$\Delta P = \rho V\left(V_1 - V_2\right) = 1.165 * 10\left(15 - 5\right)$$

$$\Delta P = 116.5\,\text{N}/\text{m}^2$$

$$F = \Delta P * A = 116.5 * \frac{\pi}{4} * 65^2 386.58\,\text{kN}$$

3- $P_{Tmax} = \dfrac{16\rho A V_1^3}{27\,2} = 3865.83\,\text{kW}$

$$C_T = 4a\left(1 - a\right) = 8/9$$

$$\lambda = \frac{C_P}{C_T} = 2/3$$

$$\lambda = \frac{R\omega}{V_1} \Rightarrow 2/3 = \frac{65\Omega}{2 * 15}$$

$$\omega = 4/13\,\text{rad}/\text{s}$$

$$P_{Tmax} = T_{max} * \Omega$$

$$T_{max} = 12563.94\,\text{kNm}$$

6.6.1.2 Blade Element Theory

The forces acting on the blade can be calculated including lift, drag acting of turbine. In this theory, the blade span is divided into segments and then forces are calculated and then total forces are calculated. Considering an infinitesimal element in the rotor, The velocity that reaches to rotor is reduced by a factor (1–a) so velocity striking is V_1 (1–a) and velocity by rotating of blade is rω(1+ a_1), as shown in Figure 6.12.

Considering the segment rotating at radius r and height dr. area = 2πr dr

The velocity that reaches to rotor is reduced by a factor (1–a), so velocity striking is V_1 (1–a) and velocity by rotating of blade is rω(1+ a_1)

Φ = flow angle
α = angle of attack

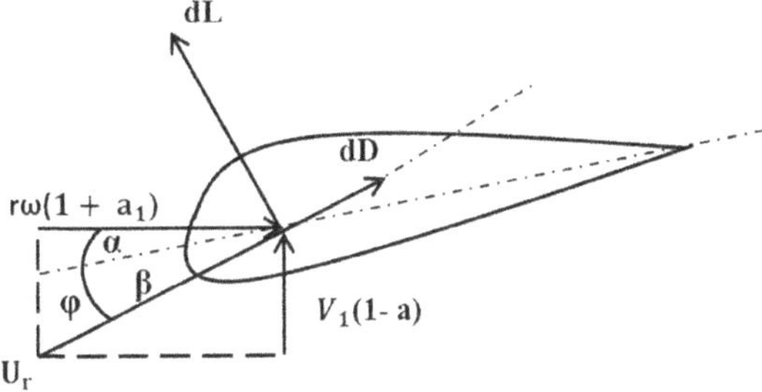

FIGURE 6.12 Infinitesimal element of rotor.

β = blade setting angle
C = chord length

$$\sin\phi = \frac{V_1(1-a)}{U_r} \;\&\; \cos\phi = \frac{\omega r(1+a_1)}{U_r} \tag{6.33}$$

$$\text{where } U_r = \left[V_1^2(1-a)^2 + r^2\omega^2(1+a_1)^2\right]^{1/2} \tag{6.34}$$

Lift and drag forces are as shown in Equations 6.35 and 6.36:

$$L = \frac{1}{2}\rho U_r^2 C C_L \tag{6.35}$$

$$D = \frac{1}{2}\rho U_r^2 C C_D \tag{6.36}$$

Forces experienced by blade at normal and tangential direction are as shown in Equations 6.37 and 6.38:

$$F_N = L\cos\phi + D\sin\phi \tag{6.37}$$

$$F_T = L\cos\phi - D\sin\phi \tag{6.38}$$

Substituting Equations 6.35 and 6.36 in Equations 6.37 and 6.38, we obtain Equations 6.39 and 6.40:

$$F_N = \frac{1}{2}\rho U_r^2 C C_L \cos\phi + \frac{1}{2}\rho U_r^2 C C_D \sin\phi \tag{6.39}$$

$$F_T = \frac{1}{2}\rho U_r^2 C C_L \sin\varphi - \frac{1}{2}\rho U_r^2 C C_D \cos\varphi \tag{6.40}$$

In order to obtain constant *C* at normal and tangent, divide as shown in Equation 6.41:

$$C_N = \frac{F_N}{\frac{1}{2}\rho U_r^2 C} \qquad C_T = \frac{F_T}{\frac{1}{2}\rho U_r^2 C} \tag{6.41}$$

Total thrust and torque on the blades can be calculated by multiplying by number of blades n, as shown in Equations 6.42 and 6.43:

$$dF = n\,F_N\,dr \tag{6.42}$$

$$dT = n\,F_T\,r\,dr \tag{6.43}$$

Substituting Equation 6.41 in Equations 6.42 and 6.43, we get Equations 6.44 and 6.45:

$$dF = n\frac{1}{2}\rho U_r^2 C\ dr\left(C_L\cos\varphi - C_D\sin\varphi\right) \tag{6.44}$$

$$dT = n\frac{1}{2}\rho U_r^2 C\ r\,dr\left(C_L\sin\varphi - C_D\cos\varphi\right) \tag{6.45}$$

6.6.1.3 Strip Theory

It is the combination of axial momentum theory and blade element theory. The force determined from both the above theories is equated to get the series of equations.

According to the axial momentum theory and blade theory, force, respectively, is

$$dF = 4a\left(1-a\right)\rho\pi r V_1^2\,dr$$

$$dF = n\frac{1}{2}\rho U_r^2 C\,dr\left(C_L\cos\phi + C_D\sin\phi\right)$$

Equating both the equation and using Equation 6.33, we get

$$\frac{a}{\left(1-a\right)} = \frac{nC}{8\pi r}\frac{\left(CL\cos\phi + CD\sin\phi\right)}{\sin^2\phi} \tag{6.46}$$

Solidarity (equation 6.47),

$$\sigma = \frac{nC}{2\pi r} \tag{6.47}$$

Expression for axial induction factor (Equation 6.48):

$$\frac{a}{(1-a)} = \frac{\sigma}{4} \frac{(CL\cos\phi + CD\sin\phi)}{\sin^2\phi} \tag{6.48}$$

Similarly equating for torque and using Equation 6.33 we get:

$$\frac{a_1}{(1-a_1)} = \frac{nC}{8\pi r} \frac{(CL\cos\phi + CD\sin\phi)}{\sin\phi\ \cos\phi} \tag{6.49}$$

$$\frac{a_1}{(1-a_1)} = \frac{\sigma}{4} \frac{(CL\cos\phi + CD\sin\phi)}{\sin\phi\ \cos\phi} \tag{6.50}$$

Power developed:

$$P = \int_0^R r\omega\, dT$$

All the above theories give an insight to the behaviour of the rotor and help in designing the blades. Several other theories and algorithms are available for defining aerodynamic properties of wind turbine.

6.6.2 Rotor Design

It is a tedious work to design a wind energy conversion system maintaining structural integrity which should be reliable, optimum and cost-efficient. The basic method to design a rotor is as discussed below. Input parameters required are as follows:

1- Number of blades (n).
2- Radius of rotor (R).
3- Angle of attack of aerofoil lift (α).
4- Design coefficient of lift (C_{LD}).
5- Tip speed ratio at design point (λ_D).

Power expected from the turbine directly depends on radius of the rotor. Power available in wind and power obtained is different; this is because various losses occur during the energy transmission or conversion. Considering power expected to be P_D, as shown in Equation 6.51:

$$P_D = \frac{1}{2} C_{PD} \eta_d \eta_g \rho A V_D^3 \tag{6.51}$$

Hence, the radius of rotor is given as

$$R = \sqrt[2]{\frac{2P_D}{C_{PD}\eta_d\eta_g\rho\pi V_D^3}} \tag{6.52}$$

where η_d and η_g are the drive train and generator efficiency, respectively. C_{PD} is design power efficiency of rotor; it ranges from 0.4 to 0.45 and combined efficiency (drive train and generator) can be taken as 0.9.

In terms of energy, the radius is calculated as

$$R = \sqrt[2]{\frac{2E_A}{\eta_O\rho T\pi V_M^3}} \tag{6.53}$$

where η_O is the overall efficiency, V_M is mean velocity and T is number of hours.

Angle of attack and design coefficient of lift are available performance data of aerofoil; if somehow it is not available, then it needs to be calculated conforming to the minimum C_D/C_L ratio which is obtained from $C_D - C_L$ graph. Draw tangent to the curve from origin to obtain coefficient of lift and drag.

Number of blades is inversely proportional to the tip speed ratio. Higher the tip speed ratio lower will be the number of blades. After obtaining all the input parameters that are identified, blade setting angle β and chord length C is calculated by following mathematical relation as shown in Equations 6.54 and 6.55:

$$\frac{\lambda_T}{r} = \frac{\lambda_D}{R} \tag{6.54}$$

$$\Phi = \frac{2}{3}\tan^{-1}\frac{1}{\lambda_T} \text{ and } \phi = \beta + \alpha$$

$$C = \frac{8\pi r}{nC_{LD}}\left(1 - \cos\phi\right) \tag{6.55}$$

6.6.3 Rotor Performance

After the completion of rotor design, it is tested in a wind tunnel, which is a scaled-down version of wind turbine. In tunnel, it is tested for various environmental conditions. One method to find out the rotor performance is by tip speed ratio, since it is the ratio of tip speed to the wind speed. In wind tunnel, the rotor behaviour is noted at various velocities which can help plotting the graph between power coefficient and tip speed ratio. Also optimum angle of attack ensures lesser power losses and reduced rotor noise. As an application to generate electricity we prefer:

1- Low solidarity.
2- Minimum number of blades.
3- Working at high tip speed ratio.

Wind turbine with one blade produces more power when compared to two blades or three blades. For aerodynamic and structural stability always three-bladed rotors are considered.

The power coefficient of the rotor is estimated as follows:

$$CP = \frac{8P}{\pi \rho d^2 V^3} \tag{6.56}$$

Example: A wind pump develops 1.6 kWh per day for irrigation; mean wind velocity at site is 4 m/s and overall efficiency is 0.14. Consider density of air 1.22 kg/m². Calculate radius of rotor.

Solution:

Given: E_A =1600 Wh ρ =1.22 kg/m² V_M = 4m/s η_O = 0.14

$$R = \sqrt[2]{\frac{2E_A}{\eta_O \rho T \pi V_M^3}}$$

$$R = \sqrt[2]{\frac{2*1600}{0.14*1.22*24*\pi 4^3}} = 1.97\text{m}$$

Example: Calculate lift coefficient, corresponding tip speed ratio, blade angle and flow angle. Radius is same as above 1.97 m, number of blades as 3, chord length and design tip speed ratio as 0.325 m and 4, respectively. Refer to the following table:

Solution:

Given: R = 1.97m C = 0.325m λ_D = 4 n = 3

Steps 1: Calculate λ_r

$$\frac{\lambda_r}{r} = \frac{\lambda_D}{R}$$

Step 2: For corresponding λ_r calculate ϕ and β

Use eqn. $\Phi = \frac{2}{3}\tan^{-1}\frac{1}{\lambda_r}$ and $\phi = \beta + \alpha$

Step 3: Calculate C_L

$$C = \frac{8\pi r}{nC_{LD}}(1 - \cos\phi)$$

By solving we get:

S. No.	r (m)	λ_r	C (m)	Φ	C_L	α	β
1	0.5	1	0.325	30	1.726	9.9	20.1
2	0.75	1.52	0.325	22.22	1.435	6.8	15.42
3	1	2.03	0.325	17.48	1.19	3.6	13.88
4	1.25	2.538	0.325	14.34	1	1.9	12.44
5	1.5	3.045	0.325	12.12	0.86	0.2	11.92

In the above example, we have a chord length constant and the blade setting angle is varying nonlinearly. If chord length is varying then angle of attack and blade setting angle can be kept constant.

6.7 FUTURE SCOPE

The wind energy industry is growing exponentially; this is reflected by their increased size from kW to MW. Now the growth of their development has reached to the offshore, which is now well established. Some companies have built and working on enhancing the capacity of floating wind turbines. Several research studies are conducted by to design bladeless turbines that will cut down the manufacturing cost, and some are focusing on putting turbines to higher altitudes and floating in sky to capture more energy. The first airborne turbine was designed by Altaeros Energies, which was installed in 2014 in Alaska at 1000 feet high and captured five times more wind power than in ground. It is buoyant airborne turbine—BAT which is like a balloon filled with helium and allows wind to pass through wind turbine. This chapter gives an outline of the different types of wind turbines; power generated from wind turbines their design and performance. Besides calculation individuals will understand various factors affecting the design of blades, which is the most important part of its designing. A sustainable energy resource has always been an attractive source and is increasing in percentage because they have low emission of greenhouse gases. Wind farms are located in an area of high wind away from bird habitat. Nowadays, offshore wind turbines are gaining interest and can be seen in many coastal areas. It is an economical method to generate electricity and beneficial for the people in rural areas to receive electricity at an affordable cost. It is one of the most promising areas in research and we still need to develop good economical turbines to meet the demand without affecting the environment.

REFERENCES

1. *Total energy consumption. Enerdata.* (n.d.). Retrieved September 17, 2021, from https://yearbook.enerdata.net/total-energy/world-consumption-statistics.html.
2. U.S. energy Information administration eia – independent statistics and analysis. History of wind power – U.S. Energy Information Administration (EIA). (n.d.). Retrieved September 13, 2021, from https://www.eia.gov/energyexplained/wind/history-of-wind-power.php

3. *Wind turbines: Upwind or downwind?* (n.d.). Retrieved September 17, 2021, from http://xn--drmstrre-64ad.dk/wp-content/wind/miller/windpower%20web/en/tour/design/updown.htm.
4. *Types of Wind Turbines.* (n.d.) Types of Wind Turbines – Energy Education, energyeducation.ca/encyclopedia/Types_of_wind_turbines.
5. *The Power of the Wind: Cube of Wind Speed,* (n.d.) xn--drmstrre-64ad.dk/wp-content/wind%20/miller/windpower%20web/en/tour/wres/enrspeed.htm.
6. Ftexploring.com, www.ftexploring.com/wind-energy/wind-power-coefficient.htm.
7. Young, J., Tian, F. B., Liu, Z., Lai, J. C., Nadim, N., & Lucey, A. D. (2020). Analysis of unsteady flow effects on the Betz limit for flapping foil power generation. *Journal of Fluid Mechanics, 902*, 30–31.
8. Manyonge, A. W., Ochieng, R. M., Onyango, F. N., & Shichikha, J. M. (2012). Mathematical modelling of wind turbine in a wind energy conversion system: Power coefficient analysis. *Applied Mathematical Sciences*, 6, 4527–4536.
9. Ragheb, M. (2014). Optimal rotor tip speed ratio. *Lecture notes of Course no. NPRE, 475.*
10. *Can Renewable Energy Sources Power the World?* OpenLearn, www.open.edu/openlearn/ocw/mod/oucontent/view.php?id=73763§ion=5.
11. Ladson, C. L., Brooks Jr, C. W., Hill, A. S., & Sproles, D. W. (1996). Computer program to obtain ordinates for NACA airfoils.
12. *AP4ATCO – Aerofoil Terminology – SKYbrary Aviation Safety.* SKYbrary Wiki., www.skybrary.aero/index.php/AP4ATCO_-_Aerofoil_Terminology.
13. Hansen, M. (2015). *Aerodynamics of wind turbines.* Routledge

7 Sustainable Bioenergy Generation Based on Agro-Industrial Residues from Paper Industries in India

Vidhi Shah, Vishweash Gurjar, Meera Karamta, and Siddharth Joshi
Pandit Deendayal Energy University, Gandhinagar, India

CONTENTS

DOI: 10.1201/b23013-7

7.1 INTRODUCTION

In the coming decades, humankind will face serious issues managing municipal solid waste. Waste generation across the globe is 2.01 billion tonnes annually with a per capita waste generation varying from 0.11 to 0.45 kg. It is projected that by the end of 2050 the global waste generation will increase with the boom of 3.40 billion tonnes per year [1]. The increment in population and growing industries are among the major reasons for solid waste generation. Taking the scenario of India, with a population of 1.38 billion it generates around 0.15 million tonnes of waste per day [2]. With the growing population, the major challenge for the Indian government is to meet the expanding energy needs of people. In an era like this, scientists are constrained to find alternative renewable energy sources that are cheap, reliable, and abundant in nature. One such unemployed source is biomass energy.

One major waste-generating sector in our country is Agro Industries. An average of 700 pulp and paper mills are functional in India currently [3]. These industries are vastly responsible for waste pollution. Additionally, paper and pulp mills are highly water-intensive in terms of operation. Although the requirement of graphic paper (newsprint, coated and uncoated paper) is declining, this industry has realigned its profit due to ever-increasing demands of tissues as well as industrial and domestic packaging requirements and other fiber-based products [4]. It is reported that the Indian paper and paper products market is expected to increase to $13.4 billion by 2024 against $8.6 billion in 2018. A compound annual growth rate (CAGR) of 7.8% during 2019–2024 is suggestive at present [3].

Of the total annual Municipal Solid Waste (MSW) generated around the globe, at least 30% is expected to be produced by pulp and paper mills. Although these industries recycle, they depend heavily on raw materials. These industries generate various types of energy-rich biomass such as forest residues, paper sludge, barks, chipper dust, etc [5]. depending upon the quality of raw materials used. In order to conserve bioenergy, the Indian government has come up with several schemes and policies to promote waste management and generate electricity from renewable sources. In this study, we propose to use the appropriate solid waste generated by the paper industry to generate bioenergy. The chapter firstly discusses biomass as a possible renewable energy source along with the techniques that could be used to derive energy from waste. And then we will see the growth of agro-industries over the decades and waste generated by the same. Lastly, we will discuss the role of the Indian government in promoting the generation of bioenergy through different schemes and policies.

7.2 BIOMASS AS POTENTIAL ENERGY GENERATION ALTERNATIVE

7.2.1 Growth of Biowaste Generation in Urban Areas

Waste generation is an inescapable part of all human activities. As India is moving toward modernization, the quantity of waste generation is rising. Along with this, the quality of waste is more diverse than in the last two decades. India contributes about 17.7% of the population worldwide, of which 35% of the population consists

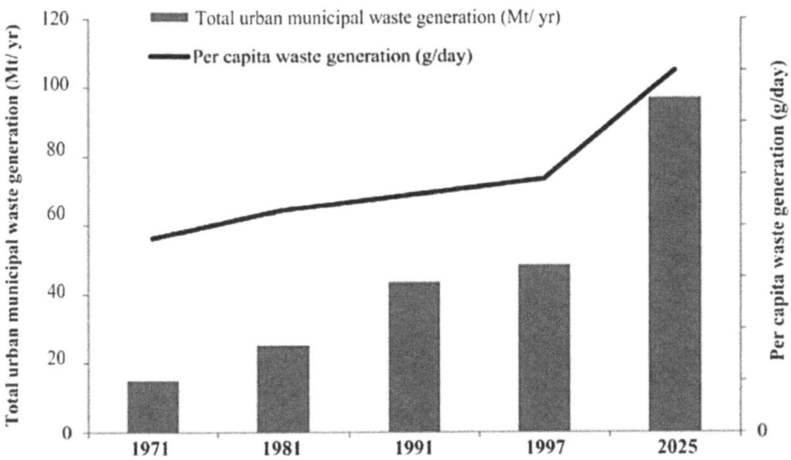

FIGURE 7.1 Growth of MSW over years.

of urban areas. Currently, 1,27,486 tonnes of waste are generated from commercial and residential activities every day [6, 7]. The increment of MSW over decades is shown in Figure 7.1.

7.2.2 Current Practices of Waste Management by Municipal Authorities

At first in India, there was not a lot of mindfulness about solid waste management and its chain of importance. But, for the last couple of years, the situation of strong waste administration has been evolving persistently. Until 1980, there was very little information accessible about solid waste management and its administration. Mostly, the solid waste generated comprises 51% of biodegradable waste. Figure 7.2 shows the typical contributors to MSW production in India [8, 9].

The total tonnes per day (TPD) of solid waste generated in some cities of India is shown in Figure 7.3. In the majority of cities, a considerable amount of collected waste is disposed of in open fields (refer Figure 7.4) [8]. Conventional methods of waste management such as mass burning and chaotic dumping have resulted in severe degradation of the environment resulting in health issues [7, 10]. Also, the convention of landfilling sites is a major problem in several cities.

7.2.3 Need for Conserving Bioenergy

The per capita waste generation in India is nearly 0.04 kg per day increasing at a rate of 1.3% per annum [11]. With the population as that of India, we can use this biomass as a source of energy generation as it is a carbon-neutral source. By doing this, the hazardous handling of solid waste can be neglected. The major ways used for the treatment of waste are composting and landfilling; the prior process can be tedious as the waste that is dumped in the fields is a mixture of different kinds, whilst landfilling

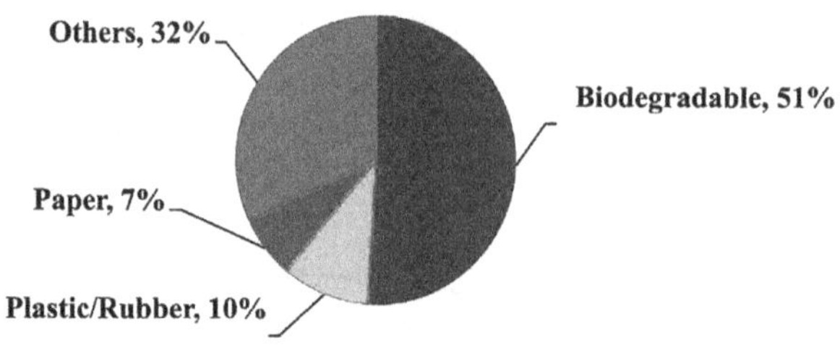

FIGURE 7.2 Major contributors of MSW production in India.

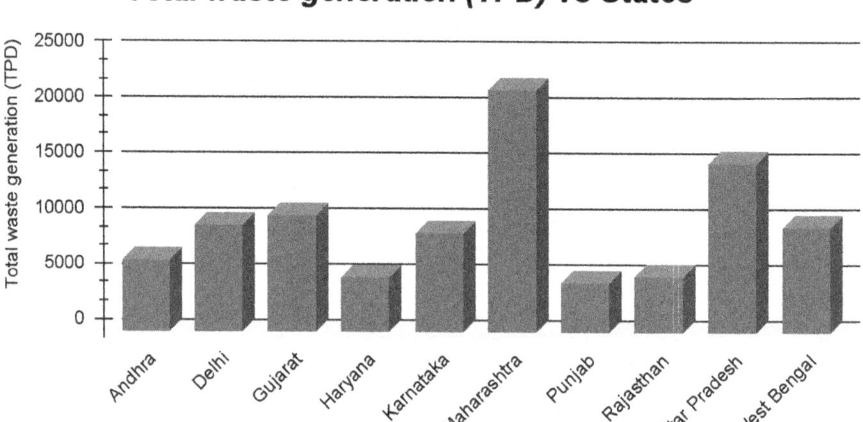

FIGURE 7.3 Total waste generation in Indian states.

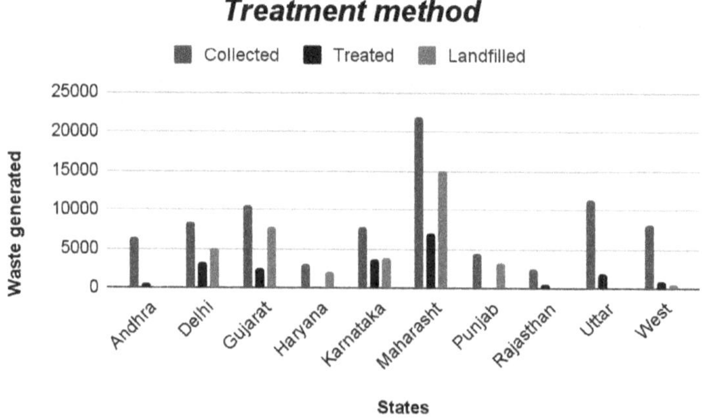

FIGURE 7.4 Treatment methods of solid waste in various states.

has higher chances of surface contamination during rains and floods. Not to forget that the entire process is not cost-effective [8, 9].

The Waste to Energy (WtE) technologies effectively treat the biowaste to generate electricity. When biowaste is burned, it generates as much electricity as the coal-fired plants do. It also removes carbon from the climate while it is delivered and returns it as it is burned. The WtE also offers observable advantages in environmental regards by offering to treat waste and substituting landfilling practices.

7.3 CURRENT SCENARIO OF AGRO-INDUSTRIES IN INDIA

India is a populous country with an average of 759 established paper mills [3]. It is one of the leading polluting industries in India, also being the most water-intensive. The growth of paper mills has increased significantly in the last two decades (as shown in Figure 7.5) [5]. A relatively enormous amount of wastewater is released into natural water bodies after primary filtration. The forecast for paper production by 2021 is projected at 521 million tonnes per annum [3]. The generated waste includes a profuse amount of wood chips and paper sludge. The generated wood pulp by the leading paper industries in India is shown in Table 7.1 [5].

7.3.1 Biomass in the Paper and Pulp Industry

Several types of biowastes are generated in paper industries at different stages of production. Industries located in different states of India have diverse contributions to solid waste. Table 7.2 shows different types of waste generated in pulp industries along with the possible ways it can be utilized and the challenges faced for the same [12, 13].

Apart from this, waste paper is also generated by household activities. These can be in the form of old newspapers, used notebooks, disposable packaging, etc.

7.3.2 Using Wastes as Fuel

Wood wastes are generated during the preparation of wood pulp in making of paper; forest residues, barks, chipper dust, pith. This waste can be dried in order to dry

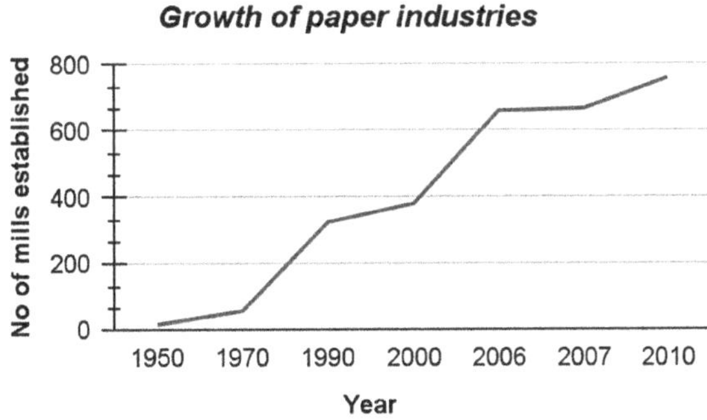

FIGURE 7.5 Growth of paper mills in last decades.

TABLE 7.1
Paper-Manufacturing Companies in India

SR	Company	Wood Pulp (t)	Wood Requirement (t) Bamboo	Wood	Total	Area Planted(ha)	Estimated Wood Generation (t) @ 60/t/ha
1	ITC Ltd.	300,000	0	1,125,000	1,125,000	140,989	8,459,340
2	Tamil Nadu Newsprint Ltd.	280,000	0	400,000	400,000	34,542	2,072,520
3	Century P & P Ltd.	280,000	168,750	958,125	1,126,875	2,340	140,400
4	JK Corp (Orissa)	220,000	50,000	850,000	900,000	59,974	3,598,440
5	JK Corp (Gujarat)	55,000	80,000	126,250	206,250		
6	Orient Paper Mill Ltd.	80,000	167,000	173,000	340,000	33,043	1,982,580
7	Star Paper Mill Ltd.	70,000	100,000	180,000	280,000	72,740	4,364,400
8	Mysore Paper Mill	60,000	0	225,000	225,000	27,500	1,650,000
9	Sirpur Paper Mill Ltd.	120,000	0	375,000	375,000	30,921	1,855,260
10	BILT Ballarpur/Asthi	220,000	200,000	175,000	375,000	36055	2163300
11	BILT Sewa	100,000	40,000	335,000	375,000		
12	BILT Yamunanagar	100,000	0	375,000	375,000		
13	BILT Kamalapur	100,000	0	375,000	375,000		
14	BILT Chowdwar (Not Operational)	0	0	0	0		
15	Seshasai P & B Ltd.	160,000	0	400,000	400,000	18,534	1,112,040
16	Andhra Pradesh Paper Mill Ltd.	220,000	180,00	800,000	818,000	124,040	7,442,400
17	Circar Paper Mill	0	0	5,000	5,000	0	0
18	West Coast Paper Mill Ltd.	280,000	0	900,000	900,000	44,260	2,655,600
19	Rama News Prints	0	0	0	0	0	0
20	HNL Kottayam	80,000	0	300,000	300,000	32,000	1,920,000
21	HNL Naogaon	100,000	0	375,000	375,000		
22	HNL Kachar	100,000	0	375,000	375,000		
23	HNL Nagaland	20,000	0	75,000	75,000		
24	Nepa Paper Mill	0	0	50,000	50,000	0	0
25	Yash Paper Mill	5,000	0	16,500	16,500	150	9,000
26	Delta Paper Mill	80,000	0	300,000	300,000	0	0
27	Emami Paper Mill	0	0	0	0	5	300
Total		3,030,000	823,750	9,268,875	10,092,625	657,093	39,425,580

the moisture content and can be co-fired along with fossil fuels to increase their caloric content. The solid waste from wood burning is completely biodegradable and non-hazardous.

Additionally, paper sludge can also be used as biowaste to generate energy. The organic fraction in paper sludge is biodegradable and its heating value is also very low. Moreover recycled and rejected paper can also be used as a bioenergy source

TABLE 7.2
Biowaste Generated along with Their Possible Utilization

Solid Waste	Source	Approximate Amount	Possible Ways of Utilization	Problems Associated with Utilization
Forest residues	Wood and bamboo cutting waste	40%	Incineration, landfill, soil-conditioning whole tree	Lower carbon and heating value, higher moisture, collection, transportation restrict its efficient utilization problem
Bark	Debarking plant	8–15%	In the manufacture of activated carbon, tannins, waxes, lignin, oxalic acid, and fiberboard, as a soil conditioner, the whole tree pulping, as fuel	Pulping wood with bark results in lower yield, higher cooking chemical, lower strength properties; some of the advantage of whole tree pulping or saving in debarking
Pith	Bagasse depicting	30% of bagasse	Can be used as fuel, cattle feed and for the manufacture of furfural, as adsorbent	
Chipper house dust	Chipper	3–5%	In the manufacture of chipboard, particleboard, vanillin building blocks, fire bricks, inferior quality of pulp, as much soil conditioner and for incineration, cooking separately or in small percentage with chips	Fines result in lower yield, higher specks, alkali carryover and chlorine consumption, lower strength, fiber pickup problem
Cyclean rejects	Pulp mill and paper section	35–50 kg per tonne of paper	In the manufacture of cheap quality board	Contamination with sand and dirt creates a problem.
Line sludge	Causticizing plant	0.45–0.65 tonne per tonne of paper	In the manufacture of portland cement, masonry cement, reburning after dislocation, soil conditioning	High alkali content causes ring formation in kilns; during the production of clinker, high moisture causes higher fuel consumption, and fineness causes jamming at various stages. Lime sludge in dry condition can be mixed with other calcareous materials. It can also be used for the manufacture of masonry cement.

(Continued)

TABLE 7.2 (CONTINUED)

Solid Waste	Source	Approximate Amount	Possible Ways of Utilization	Problems Associated with Utilization
Effluent plant sludge	Effluent plant	40–50 kg per tonne of paper	mixing with virgin pulp in small proportion, in the manufacture of the cheap quality board, as filler in the duplex board, as a soil conditioner, composting for manure	The problems are foaming, fluff and fiber pick up, lower strength, a higher percentage of higher ash percentage, Controlled mixing can overcome these problems
Coal fly ash and bottom ash	Power generation plant	0.32–0.37 tonne per tonne of coal burnt	In the manufacture of Pozzolana cement, Portland cement, brick; as a soil conditioner, adsorbent, landfill	High carbon and alkali content causes many problems in cement manufacture, low carbon fly ash is not suitable as adsorbent
Ashes from bark	Bark boiler	—	Can be used as fertilizer, lining material, landfill	
Limekiln rejects	Limekiln		In the manufacture of binder having cementitious property and as landfill	
Waste paper mixed paper, newsprint, corrugated sorted colored and white ledger, computer printout, printing presses	Municipal garbage	10–20% in municipal waste	In inferior quality board, as filler in Duplex Board	
Bagasse and rice husk fly ash	Bagasse and rice husk boiler		Both can be used as adsorbents. Bagasse fly ash of high carbon content can be used for making fine briquettes	With handling due to dust and odor and contamination with foreign material cause problem

7.4 WASTE TO ENERGY (WTE) CONVERSION

Biomass can be changed over to solid, fluid, and gases or even burned directly. It can be categorized into four types: agricultural products, alcohol fuels, solid waste, landfill and biogas. Many industries use biowastes to generate electricity; this process is called cogeneration. When biowaste is burned, it generates as much electricity as coal-fired plants. Due to the wide availability of biomass waste in industries as well as households, it is referred to as a growing renewable energy source with high growth potential.

7.4.1 Direct Techniques for Bioenergy Generation

The combustion of waste is done by extensive heating of waste in excess of oxygen. Combustion can be defined as a combination of physical and chemical processes. Firstly, the waste is air-dried and combusted with excess oxygen. As the waste is not treated with any other substance, post-combustion only ash is left which is biodegradable and can be used as manure. Combustion of MSW is a well-established process. Steam turbines are used as prime movers. The flue gases produced in the turbines are reused to produce superheated steam. Electricity is generated using steam turbines. The net plant efficiency in this process is generally between 15 and 25% [14]. The reason for low efficiency is because the calorific value of the waste is very low and not much heat can be extracted from the same. Thus, to overcome this problem we can co-fire our biowaste along with the fossil fuel to increase heat generation and thus increase the overall efficiency of the WtE plants.

7.4.2 Indirect Techniques for Bioenergy Generation

7.4.2.1 Gasification

Biomass gasification is a process that involves heat, steam and oxygen to convert biomass to hydrogen without combustion. This method converts the waste into chemical gas also known as syngas. Syngas can be combusted and passed to gas turbines for electricity generation [8, 9, 14]. Figure 7.6 shows the basic gasification method through flow charts. Air gasification is a cost-effective and widely used technology than oxygen gasification. The oxygen gasification however gives a higher heating value of syngas which increases the efficiency of this technique [14].

7.4.2.2 Anaerobic Digestion

Anaerobic digestion (AD), also known as biomethanation, is a process in which waste is decomposed by microorganisms in the absence of air [15]. Figure 7.7 represents the entire AD process. Major cities of India such as Delhi, Lucknow, and Bangalore are adopting the usage of biomethanation. Approximately 85 AD plants are at different stages of testing and operation in the country [8, 9]. Using this technique can be beneficial, as the waste generated by industries contains huge amounts of biomass. By doing the same we can make AD-based generation of electricity one of the most reliable sources of energy.

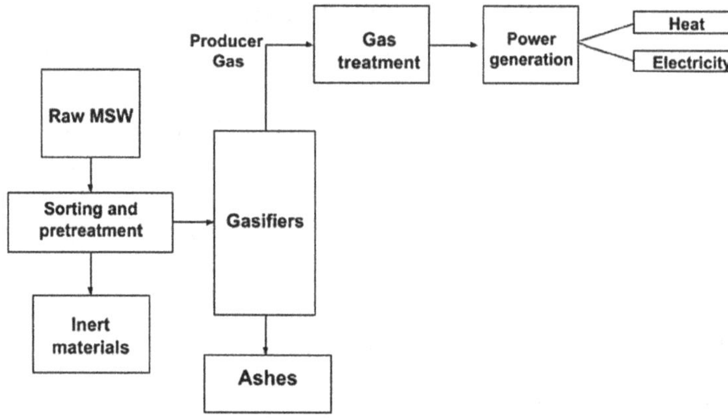

FIGURE 7.6 Flowchart of the gasification process.

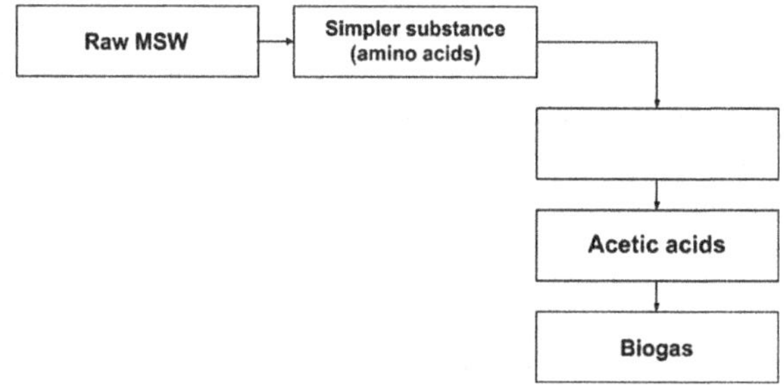

FIGURE 7.7 Schematic diagram of the AD process.

7.4.2.3 Incineration

Incineration is a method of thermal handling of biowaste generated. The waste is converted into the bi-products of CO_2 and H_2O [15]. We get ash as a residual in this process which can be easily handled and can be used as fertilizers in agriculture. The entire procedure is performed in three stages as shown in Figure 7.8.

The major challenge in establishing an incineration plant in India can be the high amount of moisture content in the waste [8].

7.4.2.4 Pyrolysis

Similar to incineration, pyrolysis is also a thermal waste conversion technique. Here the waste is treated but in the absence of oxygen [15]. The ideal temperature range for this process is 300°C–800°C [8, 9]. The waste is burned up to 300°C and then the temperature is gradually increased up to 800°C in a non-reactive atmosphere. Figure 7.9 shows a schematic diagram of the pyrolysis process.

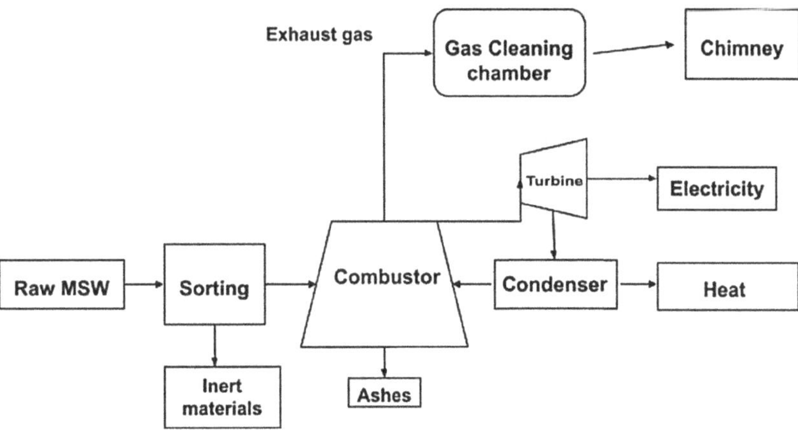

FIGURE 7.8 Schematic diagram of the incineration process.

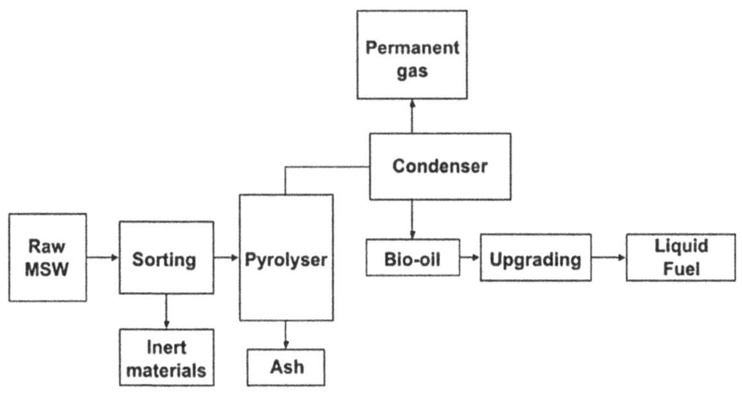

FIGURE 7.9 Schematic diagram of pyrolysis technique.

7.4.2.5 Landfilling

Landfilling the open areas with collected biowaste is the old-fashioned method of treating biomass in India. As seen in the earlier method, the production of biogas also happens in landfill technique; this gas is often called landfill gas (LFG). Because of the high humidity and high temperatures, the production of landfill gas in India is comparatively higher than in other countries across the globe [15]. Landfill gas is produced because of aerobic digestion by the microorganisms present in soil/organic matter. But, the biogas generated by landfilling has often fewer methane compositions [8, 9, 15]. The LFG is extracted through pipes, after which the gas can be processed and used as a renewable source of energy.

7.4.3 Comparative Analysis of the Techniques

See Table 7.3.

7.4.4 Different Biomass Companies in India

There are several companies in India that are using various direct and indirect technologies to treat MSW. Some of the companies along with the technology used for WtE generation are mentioned in Table 7.4 [16, 17].

7.5 POLICIES AND CURRENT TAKE OF INDIAN GOVT TO PROMOTE BIOMASS-BASED ELECTRICITY GENERATION

Indian government started various policies and initiatives to promote the bioenergy-based generation of power which are as follows: National Biofuel Policy (2018).

TABLE 7.3
Comparison of Various WtE Techniques [16, 17]

Techniques	Advantages	Disadvantages
Direct WtE process	– Produces ten times more electricity from each ton of waste – Avoids production of methane – Capable of providing baseload power	– Requires constant monitoring to make sure the emissions are safe – The maintenance cost and build-up cost is high
Gasification	– Produces lower quantity of air pollutants – Operating cost is less than a coal-fired plant – Offers wide fuel flexibility	– Frequent refueling is required for running the system – Gas resulting can clog the gasification vessel and even result in increased tar formation
Incineration	– Decreases the land size required to treat/dump waste – It has filters that trap pollutants – Prevents bad smell, as it does not accumulate waste	– Installation of the plant is expensive – Ash waste even though in a small amount can harm people as well as the environment
Pyrolysis	– Toxic components are degraded at high temperature – It is convenient and is more efficient than incineration – are flexible and easy to operate	– Requires specific amount of materials to work efficiently – Leaves toxic residues, e.g., inert mineral ash, unreformed carbon
Anaerobic digestion	– The final residue can be used as fertilizer as it is high in nutrients. – Lower production of biomass sludge – It is more cost-effective with low environmental impact	– Stabilization within reactor takes time – constant monitoring of key parameters is required – Waste generated during the process needs to be treated before discharging
Landfilling	– Easiest and cheapest concept to deal with waste – Fewer greenhouse gases emission compared to burning of waste – creates more job opportunities	– Contributes to groundwater and soil pollution – Endangerment of species may lead to ecological imbalance

TABLE 7.4
Biomass Companies in India

Company	Used Methodology for WtE Generation	Location	Installed Capacity
BMC, Kuttam	Biomass Gasification	Tamil Nadu	1.5 MW
Nuchem Ltd	Biomass gasification and co-generation	Haryana	4 MW
Indus Green BioEnergy Pvt. Ltd	Biogas based power generation	Haryana	5.6 MW
Globus Spirits Limited	Biogas based power generation	Delhi	3 MW
Hindustan Pencils	Biomass gasification	Jammu	0.4 MW
Bethmangala	Biomass gasification	Karnataka	0.5 MW

India is to fulfill fuel demand so that the country is independent of foreign imports in the oil and gas sector. The National Biogas and Manure Management Program, an initiative started by the MNRE, GoI is working on the implementation of the National Biogas and Manure Management Programme (NBMMP) across the country. It is a centralized scheme adapted by Khadi and Village Industries Commission (KVIC), State Nodal Departments and Biogas Development and Training Centres (BDTCs). Under this scheme, citizens are eligible for funds to set up biogas plants. The policies provide huge aid to the rural and semi-urban population of our country. The emphasis on biogas plant promotion is increased to help the rural population to leverage an improved lifestyle. It will ensure self-sufficiency in terms of fuel for cooking. It will also reduce the pollution created due to the burning of wood and other less efficient fuel. Access to free biomanure as a by-product from biogas plants is an added benefit for farmers and cultivators [7]. The Indian government is also currently working on a

TABLE 7.5
Various Policies to Promote Biofuel

Policies to Promote Biomass	Features
Swachh Bharat Abhiyan (Phase II)	1. Sustainable sanitation 2. Solid waste management 3. Wastewater treatment, including fecal sludge management.
JI-VAN (Jaiv Indhan- Vatavaran Anukool fasal awashesh Nivaran Yojana)	1. Financial funding of Integrated Bioethanol Projects 2. Focuses on environmental issues caused by the burning of biomass.
National Mission on environmental health and sanitation program	1. Focuses on increasing research capacity to relate climate change impact on human health.
National Urban Sanitation Policy	1. Focuses on urban sanitation. 2. Improving MSW management of the country.
National Mission on Sustainable Habitat	1. Highlights the importance and use of recycling program and strategies to circumvent greenhouse gas effects
Karnataka State policy on ISWM (Integrated Solid Waste Management)	1. Provides directions for waste management activities in a more economical and socially sustainable 2. Safeguard natural resources and public health

program for energy generation based on urban, industrial and agricultural waste; the scheme focuses on providing financial aids for projects and industries which will use biomass such as bagasse, agro-industrial waste. Initially, the scheme was introduced on a trial basis from the year 2018–2020 but was extended for another year, with reports of making it permanent [18].

Apart from these, there are several other policies [19–22] enforced by the Government of India to promote bioenergy and biomass collection. We review the salient features of these policies in Table 7.5.

7.6 CONCLUSION

In this paper, we discuss how biomass has huge untapped potential to be an alternate source for energy generation. The various technologies for the treatment of biomass produced from solid waste are discussed. In this paper, the focus is on the solid waste generated by the paper and pulp-based agro-industry. The state-of-the-art industries that deal with biomass generation and treatment in India are surveyed and reported here. A brief review of initiatives and policies that promote biomass-based electricity generation is presented. The promotion of alternate clean sources of electricity such as biomass is expected to elevate the standard of living of the rural and semi-urban population of a developing nation such as India.

REFERENCES

1. Kaza, Silpa, et al. *What a waste 2.0: a global snapshot of solid waste management to 2050*. World Bank Publications, 2018.
2. Leray, Loïc, Marlyne Sahakian, and Suren Erkman. "Understanding household food metabolism: Relating micro-level material flow analysis to consumption practices." *Journal of Cleaner Production* 125 (2016): 44–55.
3. "India's Paper & Paper Products Market 2019–2024: Applications, Raw Materials, Competition, Forecast & Opportunities", *ResearchandMarkets.com*, April 2019, https://www.researchandmarkets.com/reports/4769674/.
4. Berg, P. et al. "Pulp, paper, and packaging in the next decade: Transformational change", *Mckinsey*, August 7, 2019, https://www.mckinsey.com/industries/paper-forest-products-and-packaging/.
5. Kulkarni, H. D. "Pulp and paper industry raw material scenario-ITC plantation a case study." *IIPTA* 25.1 (2013): 79–90.
6. Bhatt, Arvind Kumar, et al. "Fuel from waste: a review on scientific solutions for waste management and environmental conservation." *Prospects of Alternative Transportation Fuels* (2018): 205–233, doi: 10.1007/978-981-10-7518-6_10.
7. Bhatia, Ravi Kant, et al. "Conversion of waste biomass into gaseous fuel: present status and challenges in India." *Bioenergy Research* 13 (2020): 1046–1068.
8. Malav, Lal Chand, et al. "A review on municipal solid waste as a renewable source for waste-to-energy projects in India: Current practices, challenges, and future opportunities." *Journal of Cleaner Production* 277 (2020): 123227.
9. Kumar, Anil, et al. "A review on biomass energy resources, potential, conversion and policy in India." *Renewable and Sustainable Energy Reviews* 45 (2015): 530–539.
10. Pujara, Yash, et al. "Review on Indian Municipal Solid Waste Management practices for reduction of environmental impacts to achieve sustainable development goals." *Journal of Environmental Management* 248 (2019): 109238.

11. Ramachandra, T. V., et al. "Municipal solid waste: Generation, composition and GHG emissions in Bangalore, India." *Renewable and Sustainable Energy Reviews* 82 (2018): 1122–1136.
12. Ashna, T. "Solid Waste Management in Pulp and Paper Industry in India", *Environmental Pollution*, November 6, 2016, https://www.environmentalpollution.in/waste-management/solid-waste-management-in-pulp-and-paper-industry-in-india/2869
13. Bajpai, Pratima. "Generation of waste in pulp and paper mills." *Management of Pulp and Paper Mill Waste*. Springer, Cham, 2015. 9–17.
14. Psomopoulos, Constantinos S., A. Bourka, and Nickolas J. Themelis. "Waste-to-energy: A review of the status and benefits in USA." *Waste Management* 29.5 (2009): 1718–1724.
15. Bhat, Rouf Ahmad, et al. "Municipal solid waste generation and current scenario of its management in India." *International Journal of Advanced Research in Science, Engineering and Technology* 7.2 (2018): 419–431.
16. "Key Indian Players in the Biomass Energy Sector", *EAI Energy Alternatives India*, Accessed 8 July 2021, https://www.eai.in/ref/ae/bio/comp/biomass_energy_companies.html
17. "Technological routes for the recovery of energy from MSW", *EAI Energy Alternatives India*, Accessed 8 July 2021, http://www.eai.in/ref/ae/wte/pro/tech/msw_energy.html
18. "Programme on Energy from Urban, Industrial, Agricultural Wastes/Residues and Municipal Solid Waste", https://mnre.gov.in/Bio%20Energy/policy-and-guidelines
19. MNRE Ministry of New and Renewable Energy, 2016. Power Generation from Municipal Solid Waste, 20th Report, Standing Committee on Energy Report. http://164.100.47.193/lsscommittee/Energy/16_Energy_20.pdf
20. MNRE Ministry of New and Renewable Energy, 2018. Government of India ministry of new and renewable energy. Generation Solid Waste.
21. MNRE Ministry of New and Renewable Energy, 2018. Government of India, New Delhi, India. Accessed 14 July 2021. http://www.mnre.gov.in/
22. MoHUA, 2016–17. Annual report. Ministry of housing and urban Affairs, Government of India. http://mohua.gov.in/upload/uploadfiles/files/annual%20report%20English%20-%2print.pdf

8 Emerging Technologies for Sustainable and Smart Cities Development

Kamakshi Rayavarapu, Anirbid Sircar, Namrata Bist, and Kriti Yadav
Pandit Deendayal Energy University, Gandhinagar, India

CONTENTS

8.1 INTRODUCTION

A smart city is an urban region that collects data using various electrical systems and sensors. Insights gathered from the data are utilised to efficiently manage assets, resources and services; in turn, the data are used to improve city

DOI: 10.1201/b23013-8

operations. Information and communication technologies (ICTs) are used in smart cities to support smarter and more sustainable development practices in areas such as transportation, water supply, and heating, as well as to promote safer urban environments.

The objective of smart cities is to be the world's top platform for sharing ideas for resolving urban difficulties so that we can live in a more resilient, sustainable, safe, and affluent environment (Neirotti et al., 2014). Smart solutions are being used in cities such as Boston, Las Vegas, Kansas City and Chicago to address transportation, sanitation, connection and safety challenges in their communities. Thomas Xu, Huawei's President of Global Government Sales, Enterprise BG, speaks with the smart cities world about the necessity to speed up the education industry's digital transformation. With new digital models appearing across the board in classrooms, research labs, and administrative offices, today's education is becoming more customised and hybrid. Fresh innovations and information communications technology are altering the education industry, and intelligence and the cloud are becoming increasingly important.

According to the census of India report, India's urban population increased from 20 % in 1971 to 34.9 % in 2020 growing at an average annual rate of 1.15%. This trend is expected to continue through 2030. But the country faces numerous challenges, including maintenance of civic amenities, drinking water, traffic regulations, high energy demand, waste disposal, high volume of rainwater during the rainy season and floods, hygiene, health and a shortage of urban houses.

The flagship urban infrastructure development scheme for satellite towns (UIDSST) system was launched for satellite towns in the vicinity of seven megacities. Pilkhuwa and Sonepat (near Delhi), Sanand (near Ahmedabad), Vasai-Virar (near Mumbai), Vikarabad (near Hyderabad), Sriperumbudur (near Chennai) and Hosakote (near Bengaluru) were among the seven towns covered by the scheme. The Scheme guidelines were aimed to help cities in developing "city development plans" (with the support of the JnNURM toolkit) and phased project planning that aligned to the funding pattern. To help cities comprehend and implement the reform agenda, a 'compendium of primers for undertaking reforms' was published.

Due to a huge substantial increase in urban population urban mega schemes was launched in the year 2015. One of the urban mega schemes is the smart city mission. A "smart city" has created a technical infrastructure that allows it to collect, aggregate and analyse real-time data to better its citizens' lives. A proper definition or a universally accepted definition of the term "smart city" has not been abbreviated. It depends on different things to different people (Roy, 2016). Indore, which has won swachh city accolades four times in a row, was named the overall smart city winner for the year 2020, along with Surat, in the Indian smart cities award contest (ISAC) of the union ministry of housing and urban affairs. "Table 8.1 shows India's urban scenario which is drastically increasing from the year 2011 to 2031."

The smart cities mission of the Indian government is an innovative and new project aimed at promoting economic growth and improving people's quality of life by promoting local development and utilising technology to produce smart solutions for residents. Smart cities are designed to make the best use of existing space and

TABLE 8.1
India's Urban Scenario

Indicators	2011	2031
Urban Population	377 Million	600 Million
Million plus Cities	53	78
Housing Shortage	18.76 Million Units	30–40 Million Units
Slum Population	95 Million	150–200 Million

Source: https://www.niua.org/.

resources while also efficiently dispersing benefits. It also aims to promote communication between residents and the administration and the broader public on a variety of levels. Smart cities are supposed to be eco-friendly. A smart city must provide a great quality of life for its residents while also fostering economic progress. This entails offering an integrated array of functionalities to citizens while reducing infrastructure costs. "Figure 8.1 depicts the reasons to need smart cities for sustainable world development."

A distinct set of frameworks is necessary for the development, integration, and application of smart city capabilities to achieve the focus opportunities for improvement and innovative technology essential to smart city projects. This ICT architecture includes an intelligent system of interconnected objects and gadgets (also known as a digital city) that sends data through wireless technologies and the cloud. Cloud-based IoT apps collect, analyse, and manage real-time data to help municipalities, organisations, and individuals make better decisions that improve quality of life (Letaifa, 2015). Citizens communicate directly with smart city biodiversity in a range of methods, including their smartphones and other mobile devices, as well as connected homes and cars. When devices and data are connected to a city's physical services and infrastructure, expenses can be minimised and sustainability managed

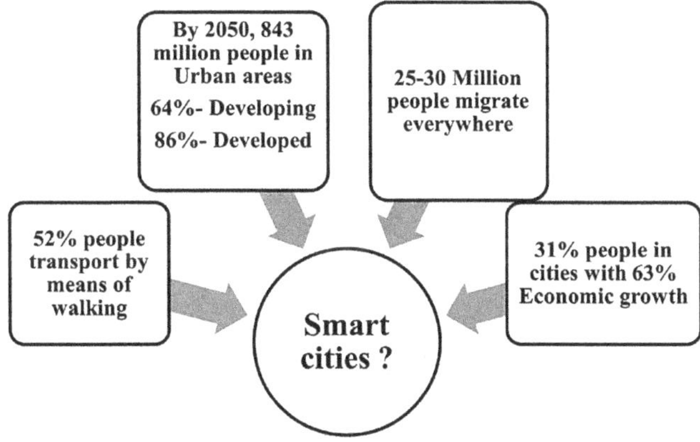

FIGURE 8.1 Reasons to need smart cities.

to improve. Communities can improve power distribution, waste management, traffic congestion, and air quality with the help of the Internet of Things (IoT).

This chapter includes the usage and working principles of smart cities, their components and characteristics. This also includes the challenges associated with smart cities and how to make smart cities operational and successful. The evolving technologies which are helpful for sustainable future energy plays an indispensable role and their advantages for future energy development are discussed.

8.2 WORKING PRINCIPLE OF SMART CITIES

A smart city is made up of numerous gadgets and systems that can make decisions based on the data they collect (Willis, 2019). This ability to make decisions distinguishes them from ordinary cities that rely on manual operation. The following are the three main steps involved in the operation of a smart city.

- Data collection is the first and most critical phase. It is carried out using a variety of devices and sensors that can be placed wherever to gather the appropriate data. This information is utilised to do further analysis to obtain the necessary information.
- Large volumes of data are evaluated to extract useful information and draw conclusions. This is usually accomplished through edge computing, which makes use of distributed systems to deliver the desired results.
- The communication step entails the wireless movement of evaluated data among servers in preparation for the next decision-making process. After final processing, the studied data is delivered to the destination, where final decisions are made.

All of the information gathered is used to make decisions. For efficient operations management, appropriate actions are done depending on the data (Lombardi et al., 2012). "*Figure 8.2 depicts the three core functions of a smart city*".

The Three Core Functions of a Smart City

Collect /Acquire data on current situations in all areas of responsibility such as power, water etc.

Analyze: Data is crushed and analyzed to generate information that can be used to improve operations and forecast what will happen next.

Communicate information to other devices, the control center, and powerful software servers

FIGURE 8.2 Core functions to operate smart city. (Modified after (https://www.twi-global.com/).)

8.3 SMART CITY COMPONENTS AND CHARACTERISTICS

Figure 8.3 depicts the smart city's components and attributes. Many components make up a smart city, and Figure 8.3 depicts eight of them. Smart cities are made up of smart buildings, smart transportation, smart infrastructure, smart energy, smart technology, smart healthcare, smart government, smart citizens and smart education.

Sustainability, life quality (QoL), economic development and smartness are all classifications of smart cities. Municipal infrastructure and governance, environment and environmental change, environmental pollution, as well as socioeconomic, and health concerns, all play potential in the evolution viability of a smart city. The emotional and financial well-being of residents can be used to determine the quality of life (QoL). Multiple factors and indications of the smart city's urbanisation include technology, infrastructure, governance and economics. The goal to improve the sustainable economic, societal and safety regulations for its citizens is defined as

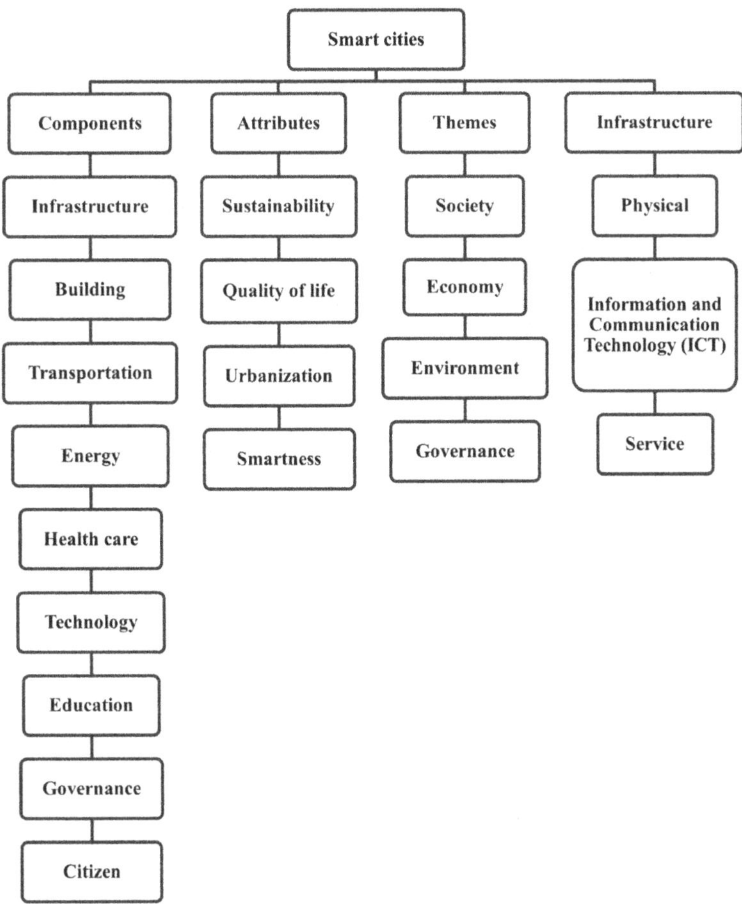

FIGURE 8.3 Smart city components and attributes. (Modified after Mohanty et al., 2016.)

smartness in a smart city. Smart people, smart economy and smart governance are some of the most typically mentioned components of city smartness (Su et al., 2011).

A smart city has four primary themes: society, economy, environment and governance. The smart city's social theme denotes that the city is for its residents or people. A smart city's economics theme denotes the city's ability to thrive in the face of ongoing job and economic growth. The smart city's environment theme denotes that the city will be able to maintain its current level of development. A smart city's governance theme suggests that the city is capable of enforcing policies and combining the other parts. Physical, information and communication technology (ICT), and service infrastructure are all part of the smart city's infrastructure. The physical infrastructure of a smart city consists of buildings, roadways, railway tracks, power supply lines and water delivery systems (kirimtat et al., 2020).

The smart city's core active component is ICT infrastructure, connecting most of the other components and functioning as the nerve centre. Physical infrastructure underpins service infrastructure, which may include certain ICT components. Smart grids and mass rapid transport systems are examples of service components (Nam & Pardo, 2011). "Figure 8.3 shows smart city components and attributes."

8.4 CHALLENGES IN SMART CITIES

A smart city is a unique way to make better use of natural resources, raise inhabitants' living standards and boost economic development. Smart city difficulties, on the other hand, must be addressed first to achieve success. The transit system in Indian cities is inadequate due to a variety of problems including a lack of investment, high population density, zoning and poor urban planning (Silva et al., 2018). To tackle this obstacle, smart city initiatives should focus on optimising mass transit use and urbanising public transportation. "Figure 8.4 depicts challenges for making smart city development."

8.4.1 Key Factors for Making Smart Cities Operational and Successful

To improve the lives of residents and to make smart cities successful the following are considered:

- Fully realising the smart city vision.
- Smart city development from a holistic perspective (rather than application-specific).
- A mindset that puts citizens first.
- Compatibility with government initiatives.
- Having a long-term perspective.
- Prioritisation of sustainability.
- Public–Private–Partnership (PPP) initiatives.

Smart cities aspire to develop modern infrastructure based on cutting-edge technologies such as the IoT, wireless communication and other innovations (Nikolov et al., 2016). Such cities are expected to maximise resource use and give the most

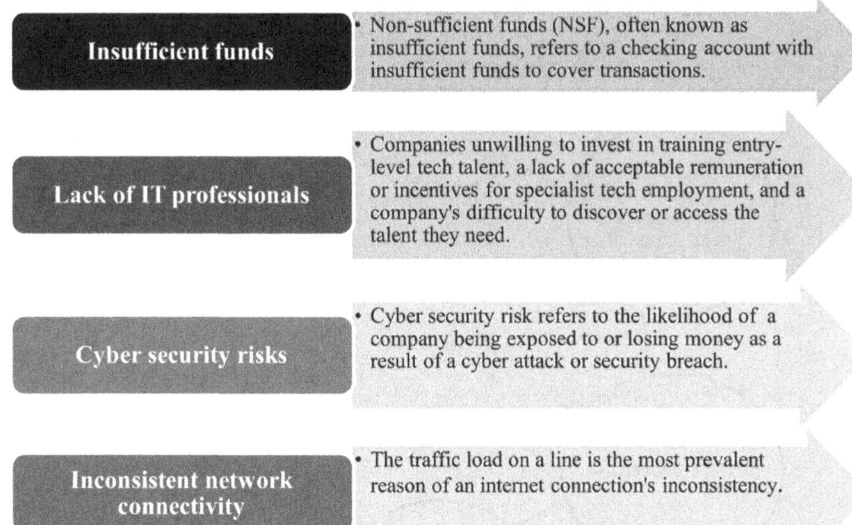

Insufficient funds
• Non-sufficient funds (NSF), often known as insufficient funds, refers to a checking account with insufficient funds to cover transactions.
Lack of IT professionals
• Companies unwilling to invest in training entry-level tech talent, a lack of acceptable remuneration or incentives for specialist tech employment, and a company's difficulty to discover or access the talent they need.
Cyber security risks
• Cyber security risk refers to the likelihood of a company being exposed to or losing money as a result of a cyber attack or security breach.
Inconsistent network connectivity
• The traffic load on a line is the most prevalent reason of an internet connection's inconsistency.

FIGURE 8.4 Major challenges to develop a smart city. (Modified after Nasution & Nasution, 2020.)

cost-effective solutions to daily issues. "Figure 8.5 shows reliable ideas for transforming the city into a smarter, safer, and more sustainable environment."

8.5 EMERGING TECHNOLOGY FRAMEWORKS

The smart city vision incorporates new technologies such as edge computing, blockchain, artificial intelligence and others to build a sustainable environment by lowering latency, bandwidth utilisation and power consumption of smart devices running diverse apps. Artificial intelligence (AI), big data and 5G are examples of technology that can improve people's quality of life significantly. Cities may create a network that provides essential urban information, such as energy measurements and weather data, by employing 5G and data, as well as deploying smart infrastructure. A smart city utilises IoT sensors, actuators and technology to connect components across the city, and it affects every layer of the city, from beneath the streets to the air that residents breathe (Ahad et al., 2020a, b). Another network framework in smart cities is big data analytics. These big data systems are effectively collected, analysed and exploited for information that may be used to improve a variety of smart city services. Decision-makers can also use big data to plan for the future growth of smart city services, resources, or places (Hashem et al., 2016). In a smart city, big data plays a vital role. Cities can discover patterns and needs by analysing data from IoT devices and sensors. Data can also help with crime reduction, smart city lighting, and water and energy systems. "Figure 8.6 shows the relationship between smart cities and big data."

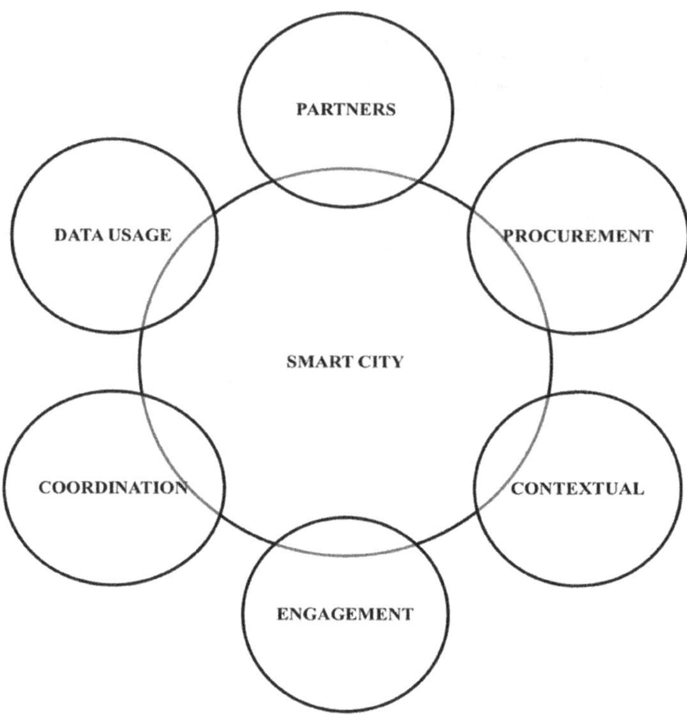

FIGURE 8.5 Reliable ideas for making the city smarter, safer and more sustainable. (Modified after (https://api.ctia.org).)

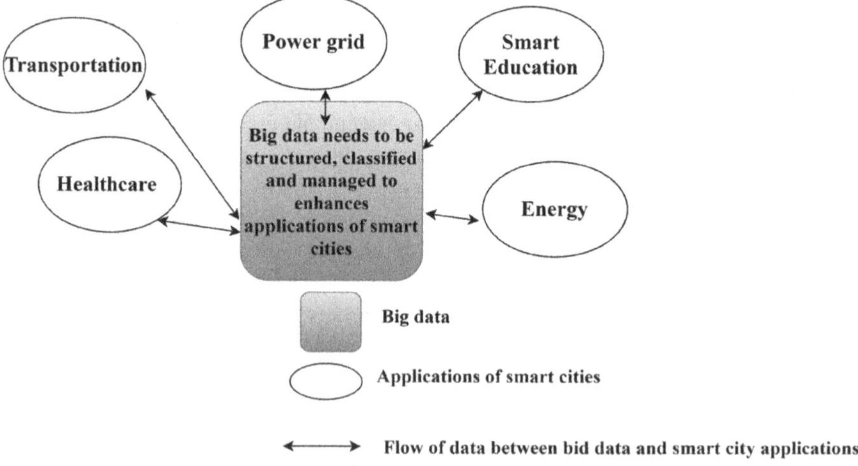

FIGURE 8.6 Smart city and big data relationship. (Modified after Al Nuaimi et al., 2015.)

The fourth industrial revolution (Industry 4.0), which focuses on advanced robotics and automation, sensor technology and data analytics, is revolutionising industries all over the world (Munasinghe & Paul, 2021). In contrast to subtractive and formative manufacturing methods, additive manufacturing (AM), often known as 3D printing, is the process of depositing material layer by layer to build a physical realisation of a 3D computer model (Iso, 2015). AM is evolving from a useful rapid prototyping technology to one that allows for complete end-product manufacture (Wong & Hernandez, 2012). This saves time, protects data and raises the quality of the final result. Blockchain, simulation, agile manufacturing, cloud, AR, robotics, and other technologies and trends are revealing new growth potential for global firms. Thus, advanced technology aids in the promotion of smart cities in a more effective manner.

In addition, The five Vs of big data management allude to some of the big data's attributes and properties.

Volume: The size of data created from all sources is referred to as volume.

Velocity: The rate at which information is evaluated, saved, analysed and processed is referred to as velocity. Recently, there has been a focus on providing Big data analysis in real-time.

Variety: The numerous database systems produced are referred to as variety. The majority of data is now unstructured and difficult to sort or tabulate.

Variability: This idea defines how the structure and meaning of data changes with time, particularly while dealing with information derived from natural language processing

Value: Refers to the potential benefit the collection, administration, and analysis of large amounts of data may provide to an organisation.

8.5.1 Internet of Things (IoT)

The IoT starts with connectivity. Because IoT is such a diverse and multifaceted field (Vijai & Sivakumar, 2016), a one-size-fits-all communication solution does not exist. "Figure 8.7 shows six of the most common wireless IoT technologies." Each solution offers benefits and drawbacks in terms of various network requirements, making it best suited for diverse IoT use cases. Each segment and application in the IoT has its own set of network needs (Lin et al., 2021). When deciding on the optimal wireless technology for an IoT application, consider issues like range, bandwidth, QoS, cybersecurity, electricity consumption, and network monitoring (Jo et al., 2019).

8.5.1.1 LPWANs

In the IoT, Low Power Wide Area Networks (LPWANs) are a relatively new concept (IoT). This technology family is designed to serve large-scale IoT networks that span extensive industrial and commercial facilities by enabling long-range networking using small, low-cost batteries that endure for years. LPWANs can connect almost any sort of IoT sensor, allowing for a variety of applications ranging from tracking inventory, environmental control, and infrastructure solutions to occupancy

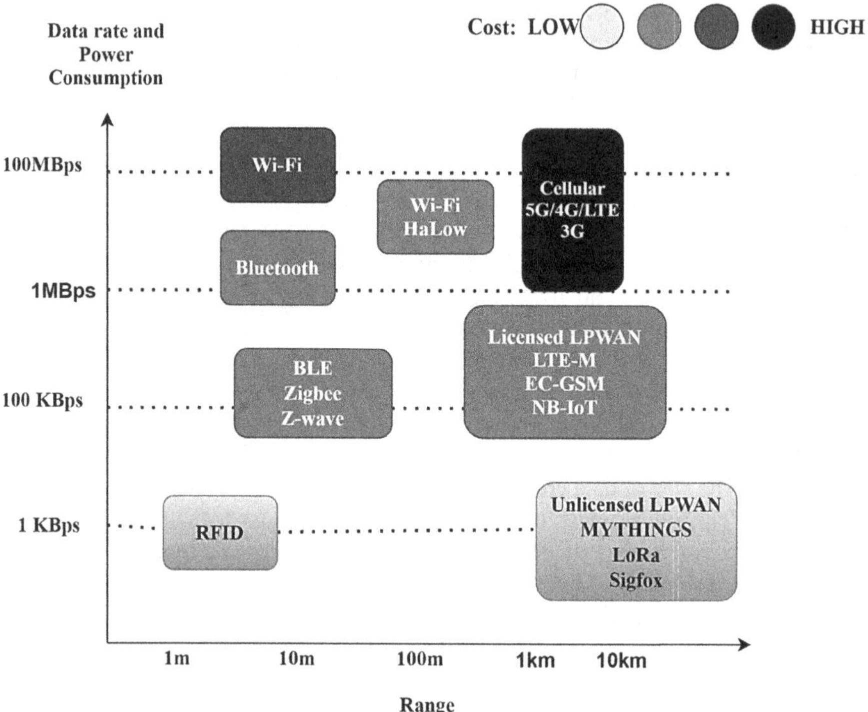

FIGURE 8.7 Wireless IoT technologies. (Modified after (https://behrtech.com).)

surveillance and consumables tracking. LPWANs can only deliver short blocks of data at a low rate, making them ideally suited to applications that don't demand a lot of bandwidth and are not time-sensitive.

8.5.1.2 Cellular (3G/4G/5G)

The majority of IoT applications rely on battery-powered sensing devices; therefore, cellular networks are not a feasible option. They are highly suited for specific use cases in transportation and logistics, such as connected automobiles and fleet management. The Internet can be used for in-car entertainment, traffic routing, advanced driver assistance systems (ADAS), fleet telematics, and tracking services. The future of autonomous vehicles and augmented reality will be cellular next-gen 5G, which will offer high-speed mobility and ultra-low latency. In the future, 5G is expected to allow for real-time video surveillance for public safety, real-time mobile distribution of medical data sets for linked health, and a variety of time-sensitive industrial automation applications.

8.5.1.3 Zigbee and Other Mesh Protocols

Zigbee is a low-power, short-range wireless technology (IEEE 802.15.4) that is frequently used in mesh topologies to improve coverage by passing sensor data across many sensor nodes. When compared to LPWAN, Zigbee offers better data throughput but lower power efficiency due to the mesh design. Zigbee and similar mesh protocols (e.g., Z-Wave, Thread, etc.) are best suited for medium-range IoT applications with an even distribution of nodes in close vicinity because of their physical short-range (100m). In most circumstances, Zigbee is an excellent companion to Wi-Fi for home automation applications such as smart lighting, HVAC controls, security, and energy management, among others – using a home sensor

8.5.1.3.1 Bluetooth and BLE

Bluetooth is a popular consumer-oriented short-range communication technology. Bluetooth Classic was designed to transfer data between consumer devices in a point-to-point or point-to-multipoint (up to seven slave nodes) format. Low-power wireless was formulated to overcome small-scale consumer IoT systems and it was optimised for power consumption. BLE-enabled devices are generally used in electronic devices, such as smartphones, which act as a hub for data transmission to the cloud. BLE is now abundantly used in the fitness and medical wearable technology (such as smartwatches, pulse oximeters, etc.), as well as Smart Home equipment (such as security systems), allowing data to be easily sent to and viewed on smart devices.

8.5.1.3.2 Wi-Fi

For both commercial and home situations, Wi-Fi is crucial in delivering high-throughput data transfer. However, in the IoT arena, the technology's fundamental constraints are in terms of coverage, flexibility and power efficiency. Wi-Fi is frequently not a viable choice for complex networks of IoT sensors that are powered by batteries particularly in industrial IoT and smart building applications, due to its high energy consumption. Instead, it refers to connecting devices that can be easily connected to a power outlet, such as smart home appliances, digital signage and security cameras.

8.5.1.3.3 RFID

Radio Frequency Identification (RFID) is a technology that uses radio waves to send small amounts of data over a short distance from an RFID tag to a reader. Technology has enabled a tremendous change in retail and logistics up to this point.

8.5.2 Smart Mobility

There are various advantages for achieving sustainable and energy-efficient mobility. The following are a few of the most important:

1. Pollution levels in the air and noise have significantly decreased.
2. Saves time by providing traffic information in real-time.
3. Improved traffic control
4. Constant connectivity, especially in the face of natural disasters and other catastrophic events.
5. Increased accessibility
6. Collaboration between small and large businesses.
7. Bettering the lives of people in rural and suburban areas.

8. It improves the economy and people's lives.
9. Allows for greater governance.

In the current ITS ecosystem, there is a potential for improvement. There can be a more secure and privacy-protected ITS with technologies like blockchain. Because of the immutable structure and hash-based coding, tampering with data safeguarded by the blockchain is extremely difficult. Every block in the blockchain is linked to the one before it, which means that if a user attempts to tamper with one block, he will have to travel the entire chain, which is almost impossible. Because ITS necessitates data gathering and integration from a wide range of devices and entities, a blockchain may be an appropriate alternative for safeguarding this information (Paiva et al., 2020). The significance of smart mobility in utilising the way we commute and travel is critical. It is estimated that the average person spends more than 35 hours per year trapped in traffic. This valuable time can be saved with the use of smart navigations and real-time traffic information. Transportation is a requirement of life; users require some form of transportation in both their personal and professional lives.

8.6 ROLE OF NETWORK MONITORING SYSTEMS IN SMART CITIES

Smart cities are ideal not only for current capacity requirements but also for providing a clear and long-term framework. This will improve smart city life in the future and should have a smart technology framework to entice and keep people. It should be able to connect and support municipal services including smart lighting systems, traffic control systems, public safety systems, electricity, video surveillance, and water infrastructure over a smart network. While this network architecture helps to overcome the digital gap among residents, it also has social benefits that might assist the whole economy. According to the Company of Motadata, 2019 IT infrastructure is so important in smart-city projects for maintaining and monitoring.

Municipalities may struggle to keep everyone connected and smart city services operational if their network management system is inefficient and underutilised. Finally, the ideal smart infrastructure is a collection of networks with ideal smart technologies and unified network management, providing visibility, decreasing expenses and providing a secure network infrastructure service and operations (Khan et al., 2020).

Benefits of NMS for Smart Cities

The glue that ties a smart city together is network management software that can interface with various network devices and allow communication to pass between systems. "Figure 8.8 depicts the role of network monitoring to develop smart cities".

Some of the key benefits of Network monitoring systems are as follows:

- Ensure data sharing across network parts and platforms.
- By allowing information to flow between systems, IT services have become more efficient.
- Log management and a proactive system have improved cybersecurity.

FIGURE 8.8 Role of network monitoring in smart city. (https://www.motadata.com/).

- Monitor all video surveillance programs to ensure public safety.
- Enhance IT service connectivity with a reliable IT infrastructure.
- Lower operational expenses with smart infrastructure monitoring and management.

8.6.1 Case Studies of Network Monitoring Solutions during Pandemic Situations

A robust solution to smart city difficulties, according to the report, should be a blend of standards and norms that include technology, processes and people. To avoid the dangers of individual-level monitoring systems, everyone must concentrate on how to use technology to change citizens' voluntary behaviour. Designing a system that builds trust between citizens and the city is the first step in changing people's behaviour to align with the greater good. Providing citizens with timely and accurate information about key topics, as well as debunking myths, goes a long way toward building trust. It enables people to comprehend which behaviours are safe and acceptable, as well as why this is beneficial to the community. Due to large gaps in education and the multiple languages are spoken, densely populated cities in nations like India confront additional hurdles. There are smart city projects in place to give inhabitants information in their native language via a Smartphone app. One of them is an AI-powered myth-debunking chatbot. Individual-level data are also important for coordinating emergency responses. Contact tracking, for example, has emerged as a critical technique in reducing the spread of the disease. Such data can be collected, analysed and reported via technology-based smart city programmes. However, data misuse erodes trust, discouraging citizens from proactively contributing their data.

Smart cities can assist every person in coping with life after COVID, but they will require both trust and technology. Singapore's government has used social media sites such as WhatsApp, Facebook, Twitter, Instagram, and Telegram to share COVID-19 information with citizens daily. Smart city technologies have already proven useful in controlling the outbreak. Here are several examples:

1. Temperature Monitoring Systems at a Distance
2. Heat maps depicting crowds in public places in real-time
3. Disinfectant-spraying drones
4. Robots serving as ambassadors from a safe distance

8.6.1.1 Temperature Monitoring Systems at a Distance

Businesses, transit networks, and public agencies are preparing to maintain normal or staggering operations during the epidemic of COVID-19. These approaches may involve an early evaluation to better identify people, infected people, and to limit the spread of COVID-19 outbreaks. Thermographic imaging and non-contact infrared thermometers are used to determine a person's temperature. Even though an infected person may be infectious without high fever or other immediately recognised symptoms, one way to confirm if someone has COVID-19 is to take their temperature (Khan et al., 2021).

Using "no-touch" or non-contact temperature assessment equipment, such as thermal imaging systems (also known as thermal imaging cameras or infrared telethermographic systems) or non-contact temperature assessment devices, is one method of measuring a person's surface temperature. Physical contact is required for other temperature-taking methods, such as oral temperature readings, which raises the risk of illness spread. "Figure 8.9 shows the thermal imaging Setup during COVID pandemic situations".

When used to take the temperature of numerous persons at the same time, thermal imaging devices are not accurate. These systems' accuracy is dependent on meticulous setup and management, as well as thorough preparation of the individual being evaluated. Several countries have deployed thermal imaging devices during epidemics, while evidence of their efficiency in preventing disease spread is equivocal. However, during a public health emergency, the FDA issued the Enforcement Policy

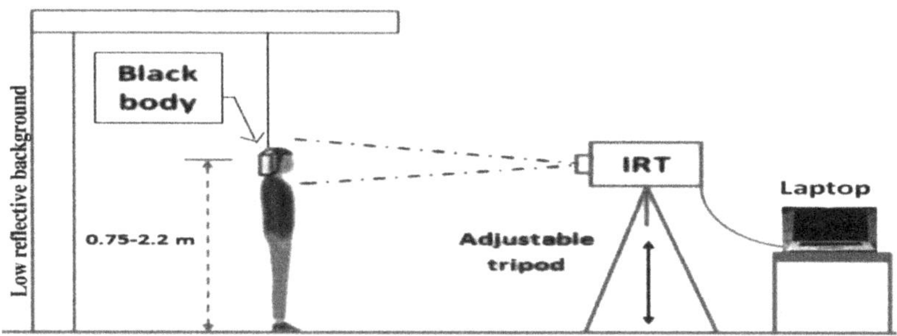

FIGURE 8.9 Illustrates to set up a thermal imaging room. (Modified after (https://www.fda.gov/).)

for Telethermographic Systems during the Coronavirus Disease 2019 (COVID-19). This Health Research Emergency guide improves the accessibility of thermal imaging devices and prevents thermometer scarcity.

8.6.1.1.1 Benefits of Temperature Monitoring Systems

It is not necessary for the person operating the thermal imaging technology to be physically near the individual being evaluated. The person in charge of the temperature imaging system may be in another area or room. The thermal imaging device may be able to monitor surface skin temperature more quickly than a traditional oral (mouth) thermometer, which requires proximity or personal contact with the individual being evaluated. Thermal imaging technologies often detect surface skin temperature reliably when used correctly, according to scientific studies.

8.6.1.1.2 Limitations of Temperature Monitoring Systems

These systems are only effective when all of the following conditions are met:

- The systems are employed in the appropriate setting.
- The devices are properly configured and operated.
- The individual being evaluated has been prepared by the instructions.
- The individual in charge of the thermal imaging system has received sufficient training.
- These devices measure the temperature of the skin's surface, which is frequently lower than the oral temperature. To account for this variation in results, thermal imaging devices must be correctly calibrated.
- It is not recommended that they be used for "mass temperature screening."

8.6.1.1.3 Use of a Thermal Imaging System

- The temperature in the room should be between 68 and 76°F (20°C and 24°C), with a relative humidity of 10–50%.
- Try to keep an eye on additional factors that could affect the temperature reading: To reduce reflected infrared radiation, avoid reflective backdrops (such as glass, mirrors, and metallic surfaces).

- Utilise a room that has no draught (air movement), is not in direct sunlight, and is not exposed to heat that radiates (for example, heaters that are portable and electrical materials).
- Prevent harsh illumination (incandescent, halogen, etc.)

8.6.1.2 Heat Maps Depicting Crowds in Public Places in Real-Time

According to The Conversation, 2020, which is an online source of a thought-provoking article, it says that heat map is a graphical representation of numerical data that use different colours to represent distinct data points within the data set. Heat maps have the advantage of converting complex numerical data into visual representations that can be comprehended at a glance. “Hot” colours indicate high user engagement, while “cold” colours indicate low user engagement. So technology has impacted the globe from a different perspective of living throughout the last 40 years. In this pandemic crisis, technology has a lot to contribute. Perhaps technology will not be able to cure COVID-19, but it will be able to stop COVID-19 from spreading at such a rapid rate. From a map perspective, the heat map solution delivers greater in-depth insights into visitor behaviours, as well as visitor flow (Khan et al., 2021). Everyone follow every consumer and gain insight from the minute they enter into the location until the moment they make a purchase using visitor flow analysis. “Figure 8.10 shows the COVID-19 heat maps helps in pandemic war.”

8.6.1.3 Disinfectant-Spraying Drones

COVID-19 facilitates the deployment of smart city technology to help cities become more resilient. As a result of the coronavirus pandemic, cities have developed drones, new sorts of monitoring, digital twins, and real-time dashboards, according to ABI research. City governments are adapting to a new reality, according to COVID-19, a worldwide technology industry advisory body, which is promoting urban resilience and digitalisation plan themes. ABI outlines the following use cases that cities have

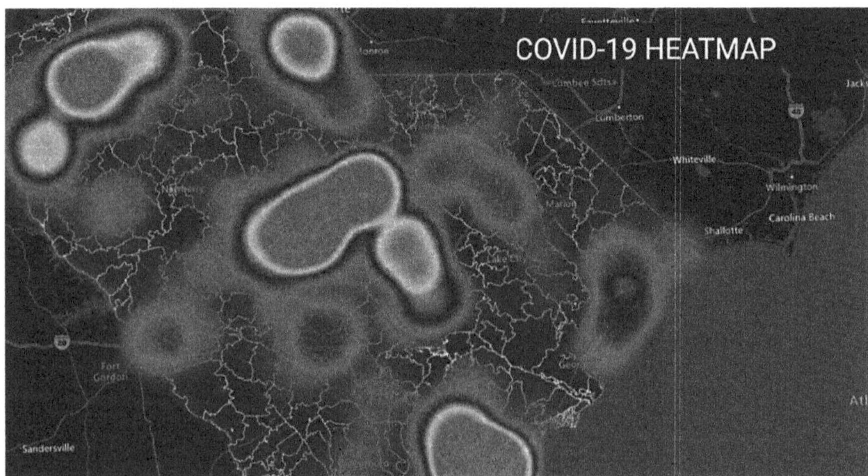

FIGURE 8.10 COVID-19 heat maps. (https://theconversation.com/).

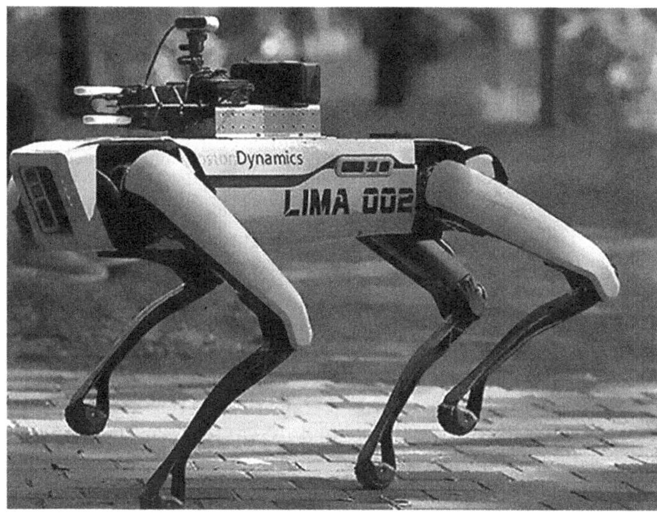

FIGURE 8.11 Following the coronavirus illness (COVID-19) epidemic in Singapore, a four-legged robot dog named SPOT patrols a park while undergoing testing to be deployed as a safe distancing ambassador. (https://economictimes.indiatimes.com).

deployed during the pandemic in its latest smart cities and smart spaces quarterly update. Digital twins provide a complete, cross-cutting, real-time view of resources, assets, and services crowdsourcing, real-time dashboards, and data sharing, including location tracking utilising Smartphone data.

Drones: Delivery of medical supplies; Rules of social distance are communicated and enforced. Surveillance in new forms: Remote temperature sensing with artificial intelligence

8.6.1.3.1 Robots Serving as Ambassadors from a Safe Distance

These are only a few of the robots that are being utilised in the COVID-19 epidemic, ranging from health care in and out of hospitals, testing automation, public safety and public works support, and continuing about daily job and life. Last-mile deliveries by autonomous vehicles (Beep, Navya, Nuro, Waymo, etc.). According to the report, cities are also benefiting from a new streaming service, as indicated by the fast adoption of e-health and teleconsultation, remote jobs, online education, e-government services and e-commerce. As a result, traffic levels have plummeted, resulting in significant reductions in congestion, fatalities, and air pollution. "Figure 8.11 shows a four-legged robot dog named SPOT patrols a park while undergoing testing to be deployed as a safe distancing ambassador."

8.7 CONCLUSION AND FUTURE SCOPE

The smart cities make use of emerging technologies to benefit their residents, but their development requires collaboration among stakeholders. One of the advantages of smart cities is the use of data to conduct municipal activities on a larger scale and more efficiently than ever before. Cities may create a network that provides essential urban information, such as energy measurements and weather data, by employing 5G and data, as well as deploying smart infrastructure. Many people have started to integrate smart city, IoT and big data to create smart city applications that will aid in achieving sustainability, enhanced resilience, effective governance, improved quality of life and intelligent management of smart city resources. These success elements, as well as a deeper knowledge of the principles, will make it possible to make a city smart. Upgrading it for smarter modelling techniques and operations will be an acceptable and long-term objective for cities to become smart. The use of new inventions and unique techniques has proven to be successful in reducing COVID-19 risk. The smart cities mission (SCM) is actively supporting COVID-19 management in terms of rapid reaction. These procedures included gathering information, quick communication, active management of COVID-infected places and persons, and taking proactive measures to prevent the pandemic from spreading. Thus, smart cities use technology and data analytics to solve some of the city's most pressing problems.

Smart energy focuses on powerful, long-term renewable energy sources that are both environmentally friendly and cost-effective. As digitisation crosses future energy systems encompassing production sites, consumers and distribution networks software is the leading technology for sustainable energy development. Thus, combining new technologies such as artificial intelligence, blockchain and edge computing reduces the latency, utilisation of bandwidths and power consumptions of smart devices running diverse apps. Data-driven smart sustainable cities are a ripe field for cross-disciplinary and transdisciplinary study, with a plethora of exciting and diverse problems awaiting academics and practitioners from a variety of city-related academic or research disciplines.

REFERENCES

(https://www.fda.gov/)

Ahad, M. A., Paiva, S., Tripathi, G., & Feroz, N. (2020a). Enabling technologies and sustainable smart cities. *Sustainable Cities and Society*, *61*, 102301.

Ahad, M. A., Tripathi, G., Zafar, S., & Doja, F. (2020b). IoT data management—Security aspects of information linkage in IoT systems. In Lianfen Huang, Sheng-Lung Peng, Souvik Pal (Eds.), *Principles of Internet of Things (IoT) Ecosystem: Insight Paradigm* (pp. 439–464). Springer, Cham.

Al Nuaimi, E., Al Neyadi, H., Mohamed, N., & Al-Jaroodi, J. (2015). Applications of big data to smart cities. *Journal of Internet Services and Applications*, *6*(1), 1–15.

Economic Times of India (2021), Availiable from https://economictimes.indiatimes.com)

Hashem, I. A. T., Chang, V., Anuar, N. B., Adewole, K., Yaqoob, I., Gani, A., … & Chiroma, H. (2016). The role of big data in smart city. *International Journal of Information Management*, *36*(5), 748–758.

https://censusindia.gov.in/

https://www.motadata.com/

https://www.twi-global.com/)

Iso, A. (2015). Standard Terminology for Additive Manufacturing–General Principles–Terminology.

Jo, J. H., Sharma, P. K., Sicato, J. C. S., & Park, J. H. (2019). Emerging technologies for sustainable smart city network security: Issues, challenges, and countermeasures. *Journal of Information Processing Systems*, *15*(4), 765–784.

Khan, H., Kushwah, K. K., Singh, S., Urkude, H., Maurya, M. R., & Sadasivuni, K. K. (2021). Smart technologies driven approaches to tackle COVID-19 pandemic: a review. *3 Biotech*, *11*(2), 1–22.

Khan, H. H., Malik, M. N., Zafar, R., Goni, F. A., Chofreh, A. G., Klemeš, J. J., & Alotaibi, Y. (2020). Challenges for sustainable smart city development: A conceptual framework. *Sustainable Development*, *28*(5), 1507–1518.

Kirimtat, A., Krejcar, O., Kertesz, A., & Tasgetiren, M. F. (2020). Future trends and current state of smart city concepts: A survey. *IEEE Access*, *8*, 86448–86467.

Letaifa, S. B. (2015). How to strategize smart cities: Revealing the SMART model. *Journal of Business Research*, *68*(7), 1414–1419.

Lin, W. L., Hsieh, C. H., Chen, T. S., Chen, J., Lee, J. L., & Chen, W. C. (2021). Apply IOT technology to practice a pandemic prevention body temperature measurement system: A case study of response measures for COVID-19. *International Journal of Distributed Sensor Networks*, *17*(5), 15501477211018126.

Lombardi, P., Giordano, S., Farouh, H., & Yousef, W. (2012). Modelling the smart city performance. *Innovation: The European Journal of Social Science Research*, *25*(2), 137–149.

Mohanty, S. P., Choppali, U., & Kougianos, E. (2016). Everything you wanted to know about smart cities: The internet of things is the backbone. *IEEE Consumer Electronics Magazine*, *5*(3), 60–70.

Munasinghe, N., & Paul, G. (2021). Radial slicing for helical-shaped advanced manufacturing applications. *The International Journal of Advanced Manufacturing Technology*, *112*(3), 1089–1100.

Nam, T., & Pardo, T. A. (2011). Conceptualizing smart city with dimensions of technology, people, and institutions. In *Proceedings of the 12th Annual International Digital Government Research Conference: Digital Government Innovation in Challenging Times*, 282–291.

Nasution, A. A., & Nasution, F. N. (2020). Smart city development strategy and it's challenges for city. In *IOP Conference Series: Earth and Environmental Science*, 562(1), 012012

Neirotti, P., De Marco, A., Cagliano, A. C., Mangano, G., & Scorrano, F. (2014). Current trends in Smart City initiatives: Some stylised facts. *Cities*, *38*, 25–36.

Nikolov, R., Shoikova, E., Krumova, M., Kovatcheva, E., Dimitrov, V., & Shikalanov, A. (2016). Learning in a smart city environment. *Journal of Communication and Computer*, *13*(7), 338–350.

Paiva, S., Ahad, M. A., Zafar, S., Tripathi, G., Khalique, A., & Hussain, I. (2020). Privacy and security challenges in smart and sustainable mobility. *SN Applied Sciences*, *2*, pp. 1–10.

Roy, S. (2016). The Smart City paradigm in India: issues and challenges of sustainability and inclusiveness. *Social Scientist*, *44*(5/6), 29–48.

Silva, B. N., Khan, M., & Han, K. (2018). Towards sustainable smart cities: A review of trends, architectures, components, and open challenges in smart cities. *Sustainable Cities and Society*, *38*, 697–713.

Su, K., Li, J., & Fu, H. (2011). Smart city and the applications. In *2011 International Conference on Electronics, Communications and Control (ICECC)*, 1028–1031, IEEE.

Town and country planning organization, Available from http://www.tcpo.gov.in/

Urban Scenario of India, Available from https://www.niua.org/

Vijai, P., & Sivakumar, P. B. (2016). Design of IoT systems and analytics in the context of smart city initiatives in India. *Procedia Computer Science*, *92*, 583–588.

Willis, K. S. (2019). Whose right to the smart city?. In Cesare Di Feliciantonio, Paolo Cardullo, Rob Kitchin (Eds.), *The Right to the Smart City*. Emerald Publishing Limited, United Kingdom.

Wong, K. V., & Hernandez, A. (2012). A review of additive manufacturing. *ISRN Mechanical Engineering* 2012: 1–10.

9 The Implication of Artificial Intelligence and Machine Learning to Incorporate Intelligence in Architecture of Smart Cities

Sayandeep Chandra and Subhankar Mazumdar
Tata Consultancy Services, India

CONTENTS

DOI: 10.1201/b23013-9

9.1 INTRODUCTION

Urbanisation is when a portion of the population shifts to the urban area from the rural sector. According to Kingsley Davis, urbanisation is a change in the ratio of the people living in the metropolitan area (Davis, 1965). Urbanisation is the result of modernisation and industrialisation. More specifically, it is a structural transformation from an agricultural society into a modern economy. This transformation happens in three stages. First, workers move from the farm field to industrial work. Second, a shift from the informal sector into the formal sector and in the last stage, there is an expansion of the urbanisation (Colmer, 2015). So, the development of the urban area's physical, social, and economic structure is an essential part of government policymaking. Also, the quality of life and living standard development are fundamental activities related to urbanisation. The establishment of the "smart cities" could direct the initiative of urban development. What could be the specific definition of the "smart city"? The answer to this general question is tricky because there is no specific statement about the definition. The word "Smart" is related to intelligence or rationality. So "smart city" could be described as the model that utilises information and technology in a realistic way that will assist sustainability and impede unplanned urbanisation (Chandra, 2019). In simple this could be concluded that smart city seeks to provide intelligent solutions by composing information and technology (Colldahl, Frey, & Kelemen, 2013). United Nations Conference on Trade and Development's (UNCTAD) report on "Smart cities and infrastructure" states that a smart, sustainable city is an innovative city that uses Information and Communication Technologies (ICTs) and other means to improve quality of life, the efficiency of urban operation and services and competitiveness. A sustainable city also ensures present and future generations concerns, economic, social and environmental aspects (Economic and Social Council, 2016). There is no fixed blueprint for developing a smart city. Plan, prepare and execute the approach in such a way that it brings a positive impact on society and its people, economically viable and environmentally sustainable. Data, analytics and IT infrastructure play a significant role in designing the solution, ensuring growth and transformation of the smart city. Big data is essential to develop insights from the enormous data set gathered from connected devices and users' information. At the same time, analytics helps the smart city optimise the operation by learning consciously and continuously from the extensive volume of data. A powerful cyber security program helps smart city authorities to protect sensitive data and digital assets from emerging cybercrimes. Security assessment and scanning over IT infrastructure help identify vulnerability and advice to adopt preventive measures against data breaches (Rolta, 2015). Adoption of digital technologies like Internet of Things (IoT), Artificial Intelligence (AI) and Machine Learning (ML) in the smart city engineering service assists in streamlining city operations and services like mobility, transportation, governance etc. In the age of technological advancement, the term "Smart City" can be rephrased as "Smart Sustainable City" which inheritates the basic characteristics of smart city with ability to manage the resources effectively. Technology is one of the driving forces employed to develop sustainability (Kubina, Šulyová, & Vodák, 2021).

The aim of the study is to explain the role of the cutting-edge digital technologies like AI, ML IoT etc., in transforming the public infrastructures and improving the operational efficiency of the smart cities. The work reviews at the high-level the implication of connected technologies to enhance the quality of life and extensively contribution towards the sustainability. The chapters are organised as follows:

- Section 9.2: Defines smart city and its attributes
- Section 9.3: Explores global trends and investment opportunities
- Section 9.4: Explains the necessity of digitalisation and policy implementation
- Section 9.5: Review of AI and ML and its broad application in developing the smart city
- Section 9.6: Throw lights on how digitalisation is shaping the smart city components like smart energy, smart building, smart transport, smart waste management and securing the system from external threats
- Section 9.7: Highlights the future scope and followed by conclusion

9.2 SMART CITY AND ITS COMPONENTS

9.2.1 Need for Smart City

The worldwide urban total population is estimated to reach 63% from 32% during 2014–2050. This trend enforces a sustainability hindrance to existing mechanisms. Around 828 million world population lived in slums in 2015. They lag basic facilities of drinking water and sanitation. This figure will increase by 6 million people per annum. It also gives rise to social instability, inequalities, unemployment, poor air quality, water scarcity, traffic congestion, urban violence, crimes, and most significant environmental pollution. Despite these adversities, cities develop various opportunities for economic empowerment. Cities facilitate around 80% of the global GDP. On average, an urban worker earns three times more than a rural worker. This relatively high amount of disposable income of urban citizens promotes their standard of living, low energy footprint on sharing common infrastructure and enables them to scale productivity levels (Estevez, Lopes, & Janowski, 2016).

In contemporary times, the existing urbanisation models can help the city grow in an innovative way where the limited resource can be utilised efficiently, balancing with development and sustainability. A technology-driven smart city deploys intelligent systems that ease socio-economic growth and improves the quality of life. The smart city scheme can abridge urban development with infrastructure and technology implementation. The leverage of information technology applications powered by the data-driven city through digitalisation (cloud computing, IoT, Open data) aids in intelligent solution creation in public services. However, it is challenging for administrators and policymakers to meet the sustainability developments by enacting the governance solution for the citizens for a better future. This need for economic growth in a proportionate momentum with sustainability has gradually changed the symbolic definitions of megacities. In 1990, around 153 million of the urban population lived ten metropolises, and by 2014 it reached 453 million in 28 megacities. A similar rise is anticipated to resume, with 41 emerging megacities to come up by 2030 (Estevez et al., 2016).

9.2.2 Prime Aspects of Developing Smart City

With the rapid urbanisation of developed and emerging economies, environmental progress is intimidating. In today's world, in many developing nations, millions of people choose to migrate to cities from rural areas with only hope for better sustenance of life. The said compelled them to live nearby and share available resources for livelihood. But this also drives pollution to expand exponentially along with the migrating population.

In the contemporary time, the cities are witnessing significant hindrances in terms of growing population, pollution, congestion, resource usage, improper infrastructures, sustainable economic growth, and stricter energy and environmental requirements. Despite all these challenges, many think tanks are optimistic about the potential for developing cities to generate less pollution during times of growth. The said can be promoted with a market creation with a proactive ecosystem of urban citizens, firms, policymakers, regulators, and non-governmental organisations to act against environmental degradation and maintain economic growth.

A typical smart city may be formed with the 'smart infrastructures' which is constituted with information systems, which can be used intelligently to optimise the utilisation of resources and improve their performance.

Currently, many efforts are being taken to prepare holistic key performance indicators for smart cities. Many urban leaders are developing innovative data streams for city performance measurement, which leverages advanced technologies and connectivity to make prompt decisions. This data is gathered through instrumentation, integrated, and analysed for intelligence on enhancing the city's services (World Development Report, 2016).

9.2.3 Classification

The urbanisation and sustainability challenges can be converged in cities around the world through the implementation of different models strategically (Table 9.1) based on the use of digital technologies in the following manner (Figure 9.1) (Estevez et al., 2016).

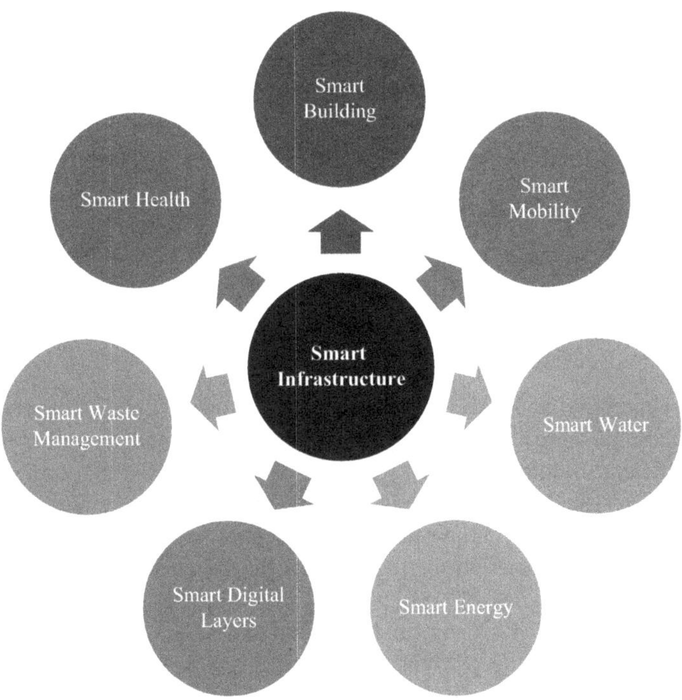

FIGURE 9.1 Smart city components (World Development Report, 2016).

TABLE 9.1
Comparing Future City Models (World Development Report, 2016)

Concept	Key Features	Examples
Digital City	• Informatics communication • City portals for online information services	Mexico City, (Mexico)
Intelligent City	• Intelligent systems functionality • Online web-based e-learning systems integrated and interoperable with other city platforms	Singapore (Singapore); Amsterdam (Netherlands); Manchester (U.K.); Helsinki (Finland); Neapolis (Cyprus)
Smart City	• Social and human concerns - the quality of life • Ecological systems -sustainability • e-Learning platform and knowledge management - Advanced visualisation and simulation tools • Benchmarking requirements	Bangalore (India); Cyberjaya (Malaysia); Konza (Kenya); Montevideo (Uruguay); Bogotá and Medellín (Colombia); Curitiba (Brazil); Barcelona (Spain); Skolkovo (Russia); New York and Seattle (USA); Hong Kong (China)
Eco City	• Natural ecosystems • Economic development while protecting the environment	Guayaquil (Ecuador); Auroville (India); Stockholm (Sweden); Freiburg (Germany); Adelaide (Australia)

9.2.3.1 Comparing Future City Models

The Smart city contributes to upgrading citizens' quality of life by promoting socio-economic growth and safeguards natural resources based on local priorities. It showcased advancements in the continuous transformation process through the engagement of stakeholders, collaboration with humans, institutional and technical capacities in a unified manner (Estevez et al., 2016).

9.3 CONSTITUTING SMART CITIES

On architecting a smart city, there are two possible ways through which it can be achieved:

- A technology-intensive city that connects the town with a lot of intelligent devices that store a lot of real-time data information and prescribes a smart or efficient solution to the given context
- Propagates the bonding citizens and governments – where the public services provided are optimised through technology platforms, employing bidirectional information sharing to better the systems.

These approaches are distinct and can be applied to develop cities to enhance public services (Muente-Kunigami & Mulas, 2015)

9.3.1 Benefits of Efficient Urban Planning and Smart City Implementation

A key differentiator between a smart city and an old city is efficiencies of city operations and management, which includes social inclusion, accountability, citizen empowerment, smarter decisions irrespective of its complexities. Better efficiency can be achieved by intelligent use of data, through which real-time analysis and solutions can be predicted to emerging problems or situations. The inclusive nature can be achieved through erecting a comprehensive geographic database of socio-economic and physical indicators for its citizens. Innovative ideas are the supply line for exploring solutions with smart systems of a complex city, which must work on a 'public-private-people-partnership' approach to other citywide scales up (World Development Report, 2016).

Cities hold a more significant portion of the population and can mitigate the adverse effects of climate change. In the emerging economies, around 60% of the area is to be urbanised by 2030 following the economic development. A city has to build as 'climate resilient' and trade-off with economies of scale and sustainability. So, a smart city must be capable of urbanising with green energy sources, energy access, a green fleet system, and green infrastructures to curb pollutions and GHG emissions potentially. The initial step for cities to develop resilience, counter climate risk, and explore climate opportunities is to assess various climate impacts. The urban planning and infrastructure investments will shape the direction of urban growth trajectory and development for future decades. Globally, 210 cities already adopted a climate change action plan, and 111 cities are in ratification to create the same (International Finance Corporation, 2018).

9.3.2 Global Trends in Smart Cities

9.3.2.1 Developed Countries 'Smart City' Approach

The IMD-SUTD Smart City Index (SCI) assesses citizens' perceptions on issues associated with structures and available technology applications in their city. The perception of citizens is based on health and safety, mobility, activities, opportunities, and governance (IMD World Competitiveness Center, 2020). According to their analysis on top 10 smart cities of Singapore, Helsinki, Zurich, Auckland, Oslo, Copenhagen, Geneva, Taipei City, Amsterdam and New York are based on the sector-specific technology solution. Most commonly applicable solutions pillars are mentioned below.

9.3.2.2 Global Investment Potential for Smart Cities

Smart cities can allure global investment opportunities for implementing measures on climate actions. To develop an ecosystem that paves a way amending climate policies, roadmaps, setting up targets, proactive actions, and achieving goals as per strategic planning. IFC expects the cumulative investment to reach US$ 29.4 trillion by 2030 across six sectors in emerging market cities (International Finance Corporation, 2018. The sectorial forecast for investment by 2030 is elucidated below.

9.3.2.3 Climate-Resilient Smart Cities

The idea of a climate-smart city is conceived by Innovate4Climate – I4C (a subsidiary of the World Bank Group) (Table 9.2), which promotes sustainability through

TABLE 9.2
Top 10 Smart City Solution Pillars

Sector	Technology Solutions
Health and Safety	• Website or App for residents to give away unwanted things • Website or App monitor air pollution • Free public Wi-Fi services • CCTV cameras for safety • Online based medical services • Digitally generated reports for maintenance issues
Mobility	• App-based car-sharing services congestion abatement • Smart parking service • App-based Bicycle hiring system in the city • Online ticket for public transport • Mobile-based traffic congestion
Activities	• Online system for bookings tickets to public shows and places
Opportunities (Work and School)	• Online access work opening and job banks • IT skills in schools curriculum • Transparent online services for business opportunities • Internet connectivity and reliability
Governance	• Online public services in finance promote corruption-free • Online voting services • Online feedback mechanism for citizens • Digital systems fasten the documentation process for service management

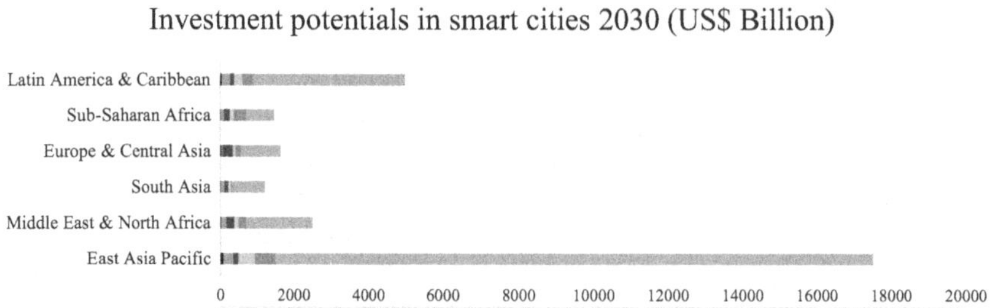

Investment potentials in smart cities 2030 (US$ Billion)

Latin America & Caribbean
Sub-Saharan Africa
Europe & Central Asia
South Asia
Middle East & North Africa
East Asia Pacific

0 2000 4000 6000 8000 10000 12000 14000 16000 18000 20000

	East Asia Pacific	Middle East & North Africa	South Asia	Europe & Central Asia	Sub-Saharan Africa	Latin America & Caribbean
■ Waste	82	22	17	28	13	37
■ Renewable energy	266	141	88	31	89	226
■ Public transportation	135	217	116	281	159	109
■ Ciimate-smart water	461	110	64	79	101	228
■ Electric vehicles	569	214	46	133	344	285
■ Green buildings	16000	1800	881	1100	768	4100

■ Waste ■ Renewable energy ■ Public transportation ■ Climate-smart water ■ Electric vehicles ■ Green buildings

FIGURE 9.2 Investment potential in smart cities.

Around rise in 2.5 billion additional people to cities, and share of urban population to escalate from 55% to 68% (2014–2050)

Currently cities are accounted for 70% of all global CO2 emissions each year (approx 25 billion tons) Asia's cities consume 80% of the region's energy and create 75% of its CO2 emissions

Almost half a billion urban residents live in coastal areas and four out of every five people impacted by sea-level rise by 2050 will live in East or South East Asia.

India and China are two countries that will account for 35% of the world's forecasted urban population growth by 2050.

Climate investment opportunities for global emerging cities are $29.4 trillion by 2030 (IFC)

Singapore examplified being 3rd highest population density-but self-resilient on climate-smart development (waste disposal design, offshore floating solar, saving of 2,000 tons of GHG emissions per year)

FIGURE 9.3 The critical outcomes.

Carbon Neutrality, Recycling, Green Catering, and Carbon Mobility (Figure 9.2). The third edition of I4C took place in Singapore (Innovate4Climate, 2019), and some of the critical outcomes of the event were (Figure 9.3).

9.3.2.4 Converging Governance Needs

In 2020, when the world is struggling to cope with the Covid-19 surge in the number of cases, Singapore came up with an innovative digital solution (Figure 9.4).

Smart road map

- Setting a path towards a smart city model for the next five to ten years.
- Developing an action plan and investments road map are proposed, based on city's ongoing context.

Prioritization of city's need

- Engagement of stakeholders: Developers, Civil organizations, local and citizens/representative body, IT professionals, public officials, urban sector specialists
- Identification of needs and priorities, anaysis on available resources, and available technology standards.
- Compare and implement industry best practices

Creating sustainable solution

- Brainstorming sessions of all stakeholders to develope a solution thourgh creative thinking, innovation and entrepreneurship.
- Partnerships with thinktanks/academia and private sector incorporation may aid exploration of apt solution.
- Development of prototypes and simulations to access the erection and impact assessment (feedback system)

Urban Innovation & Research

- Interaction between all stakeholders to nurture disruption ideas and solutions to be tested in a fail-safe environment.
- Innovation research facilities for service improvement and support stakeholders with aims on quality of life.

Networking cities

- Bind with a network to share applications and practices, so that it maximizes the value of the solutions to all public (e.g. European Network of Living Labs or the Open Cities initiative)

FIGURE 9.4 Smart city framework.

It introduced a technology-based efficient contact tracking system to combat the pandemic situation. The citizens were needed to scan QR codes for identification purposes to keep track of movements for avoiding the virus. An application was launched with an encrypted ID with digital devices that alert to shun proximity of the people in public places. Similar data-driven mechanisms were useful for curbing crime rates in the city. Singapore cites that emerging technology's ethical and wise utilisation can become a boon for our society and overcome governance challenges. Soon cities are looking to use 'Fourth Industrial Revolution (4IR)' tools to address governance challenges by incorporating disruptive features of artificial intelligence (AI), robotics, Internet of Things (IoT), and quantum computing as a suitable resolution (Antunes, Tanaka, & Merritt, 2021).

9.3.2.5 Developing Smart City Framework

On developing a smart city, the following frames are to be used to construct a smart city and promote its sustenance (Figure 9.5).

The framework for developing a smart city amalgamates technology adoption and proactive citizen's interest with the government to make a lucid and transparent public

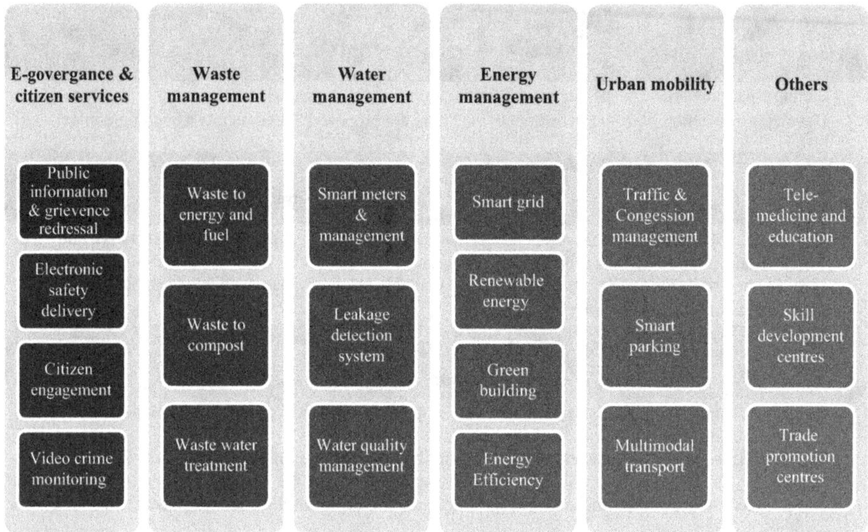

FIGURE 9.5 Smart city solution areas.

facility and service delivery mechanism. It includes a long-term technology investment plan that intelligently leverages the technology. The citizens' feedback is instrumentally significant for the government facilitated public services for the constant scope of improvement and solidifying solutions (Muente-Kunigami & Mulas, 2015).

9.3.2.6 India's Perspective of Smart Cities

India started its journey with the "Smart Cities Mission" in 2015. The prime objective of this mission is to improve the city's core infrastructure, promote a sustainability and improve the quality of life by integrating 'smart solutions'. The Government of India (GoI) pursued an ambitious goal of converging 100 old cities into smarter ones. To fuel the initiative, GoI funded the mission through a Centrally Sponsored Scheme that supports Rs. Ninety-nine cities have proposed 48,000 crores over five years and Rs.2,01,981 crore under their smart city plans (Ministry of Urban Development, 2015).

City inhabitants and their aspirations are used to prepare the Smart City Proposals (SCPs). Currently, the aggregated number of proposals at the national level is more than 5,000 projects. The mission implementation is being carried out by a Special Purpose Vehicle (SPV). SPVs are set up at the city level, acted as a limited company (approved by Companies Act, 2013), and co-promoted by the State governments and the Urban Local Body (ULB) jointly as 50:50 equity shareholders. SPVs develop the proposal, Detailed Project Reports (DPRs), conduct tenders, and implement projects or smart solutions (Ministry of Housing and Urban Affairs, Government of India, 2015).

The efficacy of' Smart Cities Mission' is purely based on overcoming challenges with a practical, sustainable solution. Most of the solutions establish core infrastructure that encompasses the following areas (Figure 9.6) (Ministry of Urban Development, 2015).

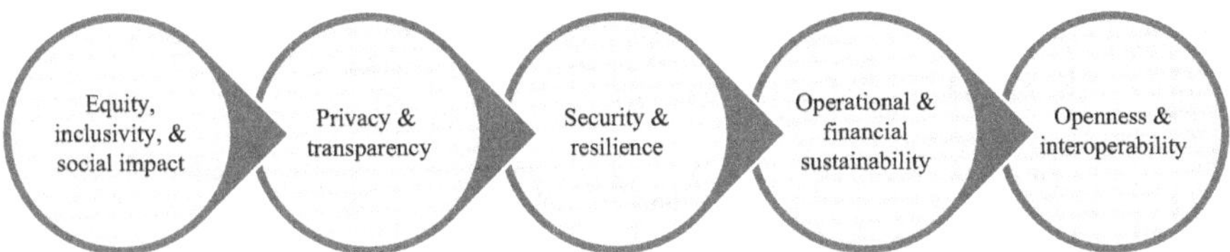

FIGURE 9.6 Governance framework.

9.4 ROLE OF DIGITAL TECHNOLOGIES IN SMART CITY BURGEON

9.4.1 A Governance Roadmap

Considering the global dynamics and transitions in economies across all sectors, United Nations set a Sustainable Development Goal (SDG-11), i.e., 'sustainable cities and communities, to make cities safe, resilient and sustainable and eventually upgrade living standards. In the contemporary time, governance holds a key for the upliftment of values city in terms of ethical and responsible smart city development. It can be achieved by deploying advanced digital technologies as a medium to connect government bodies or municipalities with public trust and feedback systems. Here some of the most commonly encountered hindrances are maintaining data privacy and protection, cybersecurity, and networkability as barriers for smart city programmes (Nations, 2015).

Though all the cities are unique and their needs different, they require a technology governance framework to be retrofitted per the local context. A typical policy roadmap must fulfil the below five principles to progress towards a 'smart city' (Antunes et al., 2021).

9.4.2 Abridging Gaps

The policy roadmap model's successful execution requires the digital system to operate comprehensively. For ensuring that, gaps in the system must be connected for effective deployment. The said involves the engagement of stakeholders and their cooperation in every manner. The local government or authorities need to adhere to the policies set by regulators at the national level. A robust governance model involves – city leaders, policymakers, civil society, smart city technology vendors, urban innovators, technology solution providers, administrators, donors, and citizens to rationalize the policies and speed up the adoption progression.

9.4.3 Policy Benchmarking

Smart city technologies have a crucial role in improving the resilience of cities for future challenges. In the same context, policy benchmarks hold a key for better smart city governance to guarantee data infrastructure's ethical and responsible development. Intelligent management can help cities recover from the COVID-19 crisis by avoiding exposing citizens to potential risks through a tracking and alert system (Figure 9.7) (Antunes et al., 2021).

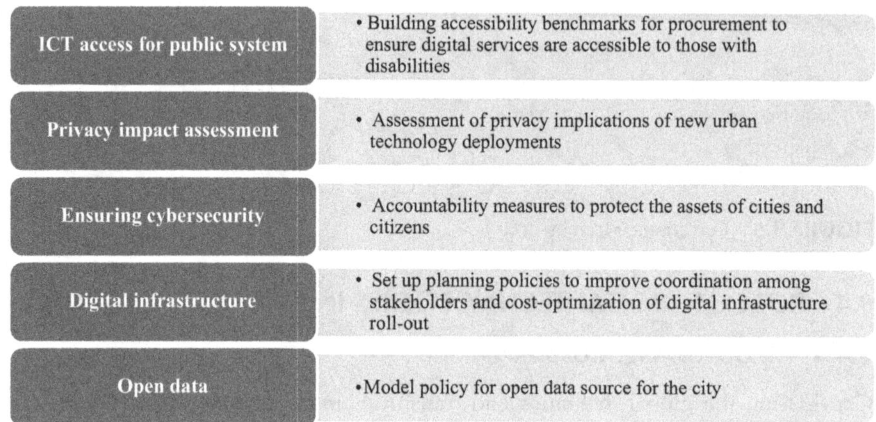

FIGURE 9.7 Data infrastructure of smart city.

A strong roadmap, complemented by benchmarking policies powered by advanced digital technologies, can make the public system more efficient, scalable, and equitable (Jeff Merritt, 2021). The benchmarking for innovative governance can be elucidated briefly keeping because of data infrastructure.

The policies for the cities are assessed against the roadmap and action plans. Responses are accumulated to form a dataset. These datasets are critically analysed, and efficiency is tested. On found satisfactory, it can be presumed as the benchmark by other cities as guidelines to converge towards a smarter one.

9.5 ARTIFICIAL INTELLIGENCE: SHAPING THE WORD

The rapid expansion of digital technology is restructuring the workflow and work dynamics across the world. The broad implication of digitalisation has made the production process, distribution and communication process smooth and helped track consumer data, social behaviour and the way people interact. Technological advancements are utilised for managing cities effectively and modernising the urban lifestyle. Emerging technologies- Internet of Things (IoT), Big data, Artificial Intelligence (AI), robotics applications, and 5G network drive the endeavour to digitalise the city life and processes. Data derived from numerous sources are analysed to understand the pattern among stages of the workflow and value chain, manage the threat related to computation or human activities, and measure the resource consumptions. AI and robotic applications have become an integral part of smart cities that enhance decision making and assist in different situations. AI is extensively used for process control, medical and healthcare industry, public safety, transportation, media and communication industry and other sectors; the adoption and judicial use of AI helps structure the smart cities.

AI is a grouping of technologies that lets machines adopt human capabilities, and it is powered by intelligent algorithms and data sets to replicate human cognitive abilities. AI enables the device to sense the environment by acquiring raw data, comprehending the insight by analysing the data, acting based on the comprehension, and finally learning to optimise the performance (Gebre & Whitehouse, 2018). The ability to learn is the fundamental concept of AI. Also, it can decide the action required to accomplish a task by analysing the available data. On the other hand, machine learning (ML) technology qualifies a machine to learn the processes by identifying the data pattern and doing the job. AI is the core part of ML, and it assists in understanding the visual inputs and information retrieval (The Royal Society, 2017). The adoption of intelligent devices, extensive social media platforms and connected devices generate big data every day. Such comprehensive and complex data are required to be analysed with accuracy to derive patterns. AI, ML and deep learning (DL) are considered the best possible digital tool to reach an optimal decision (Allam & Zaynah Dhunny, 2019).

Technology is contextualised in the smart cities for designing a convenient system and develop an affluent lifestyle for its people. A smart city should connect people, information, and city infrastructure using cutting-edge technology to establish sustainable green towns, increase the quality of life and enhance commercial aspects (Lee, Hancock, & Hu, 2014). Emphasise the development of infrastructure and adoption of smart computing nourish the growth of smart cities. The application of smart computing technology will help streamline the administrative service, make education easily accessible, improve the health care system, public safety, and optimise resource usage (Washburn & Sindhu, 2010).

Cities worldwide are metamorphic to digitalisation by implementing sensors, adopting computational cores, and faster networking systems. This complex system generates enormous environmental and human behaviour data from the specific neighbourhood (Alvarez, 2017). IoT and connected devices are the primary producers of this data. Accuracy in analysing data will pave the path for growth and transformation.

9.5.1 Machine Learning

Machine learning is one of the critical components of the field of data science. With the help of statistical methods, algorithms are trained to classify data, predict models, derive insights, and imitate human learning (Figure 9.8). ML methods are

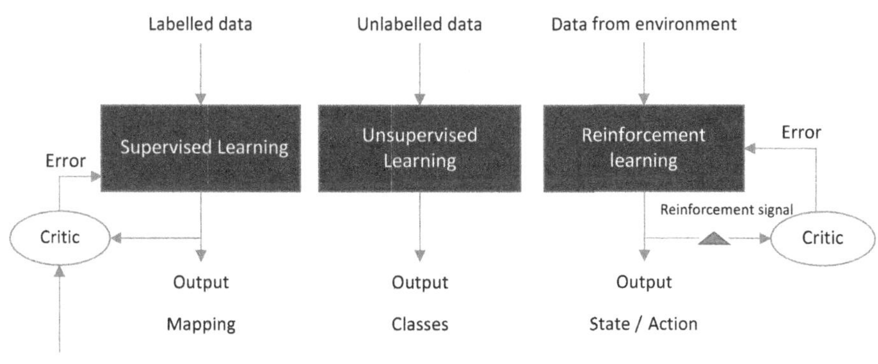

FIGURE 9.8 Learning model for algorithm.

classified into supervised learning, unsupervised learning, and reinforcement learning (RL) (Ullah, Al-Turjman, Mostarda, & Gagliardi, 2020). Supervised learning uses labelled datasets to train the algorithms that predict the outcome accurately. The data is fed into the model, and it adjusts weight till the model is appropriate. The supervised learning method applies cross-validation techniques to avoid overfitting or underfitting. In contrast, unsupervised learning analyses the unlabelled dataset to discover hidden patterns and distributions. The difference between supervised and unsupervised learning is the type of dataset that is fed into the model. Also, unsupervised learning requires human intervention to validate the output (IBM Cloud Learning, 2020). RL is a behavioural learning method where the model is trained with the trial-and-error method. Successful outcomes are used to develop the best recommendation model. In RL, there are four fundamental components- agents, environment, incentive and behaviour. The agent gathers data and the environmental condition to determine the performance behaviour. During this learning process, the algorithm explores state-action pairs within the environment, and this is using the state-action pairs, the algorithm selects optimal action for a specific goal. Q-Learning is an approach of RL to reward in an environment in stages (Abdulqadir & Abdulazeez, 2021; Jones, 2017).

Deep learning is a sub-field of ML. Deep learning mimics the neural function of human beings to think and solve a problem. The deep learning technique monitors how the human brain works and prepares algorithms for machines to resolve an issue like a human. Deep learning makes a cluster of data and gathers all the outputs from all the sets to enhance results. Compared to ML, Deep Learning takes more time to train data and can categorise the unstructured data according to the hierarchy of features. It does not require human involvement and can analyse a large volume of data (IBM Cloud Education, 2020a; Khanna, 2019).

9.5.2 Artificial Intelligence

Artificial Intelligence is the science to develop intelligent programming that is capable of understanding and imitating of human intelligence (McCarthy, 2004). AI makes it possible to analyse the complex data set to derive a solution to a problem. Machines are designed intelligently to interpret the pattern hidden in data using the algorithm (Figure 9.9). AI reinforces "experience-driven learning." to prepare a model capable of predicting, taking action, and learning from the outcome (Olayode, Tartibu, & Okwu, 2020). In simpler words, AI makes computer systems proficient in problem-solving and decision-making by analysing data. Machine learning and deep learning are two primary sub-fields of AI. The said sub-sections leverage the power of AI to compose a self-efficient system that can predict or classify, leveraging insights from data.

There are two types of AI-Weak or Narrow AI and Strong AI. Weak or Narrow AI or Artificial Narrow Intelligence (ANI) can perform only the task it is designed for. ANI is better to perform functions that are made for humans- learn about the problem and solve it (Duin & Bakhshi, 2018). On the other hand, Strong AI is a conceptual form, aims to design machines with super-intelligence that can think like individuals - self-aware, solve a problem, learn and think for the future (IBM Cloud Education, 2020a, 2020b). There is no practical example of Storng AI.

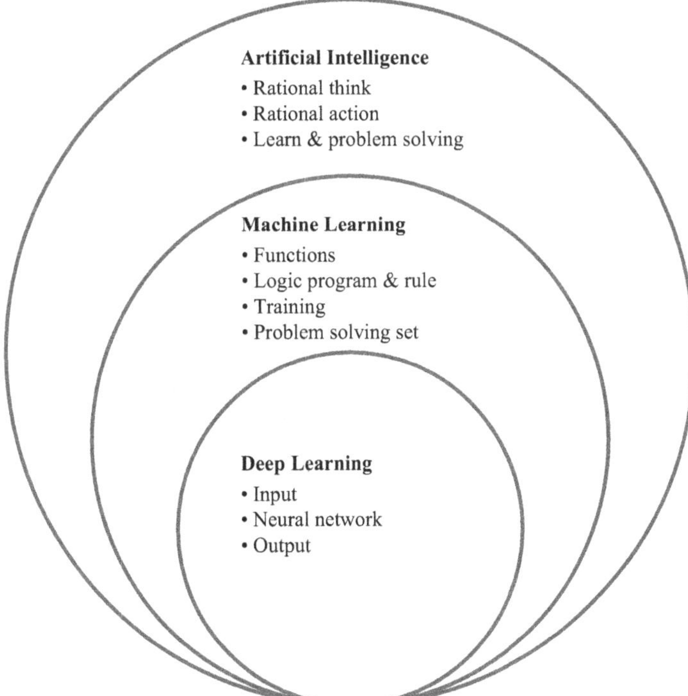

FIGURE 9.9 Relationship among AI, ML and deep learning.

9.6 DIGITALISATION OF SMART CITIES

A Smart City orderly implements digital technologies to decrease resource input, improve quality of life, and enhance a sustainable economy. It requires the use of intelligent solutions for infrastructure, energy, housing, mobility, services, and security based on connected sensor technology, connectivity, data analytics, and other value-added processes. Smart city technologies assist in streamlining operations across city services like transportation, safety, e-governance, health etc. Digital technologies applied for Smart cities are:

- Smart-Energy management: Smart energy management system analyses the energy requirements and their sources. Based upon the insights optimal mix of the sources for the energy generation can be designed, and organised and secure distribution system can be developed, and monitoring of the consumption pattern is possible. The said solution efficiently manages the carbon footprint by considering the environmental aspect as a driving factor.
- Smart-Transportation: Smart-Transportation includes end-to-end transport solutions for vehicle tracking, monitoring, parking space management, passenger information, drivers' behaviour analysis and optimisation of fuel consumption and many crucial aspects.

- Smart-Building: Smart buildings are sustainable and secure while monitored and managed by digital features. Integrated solutions help in optimising energy consumption and explore alternative energy resources.
- Smart-Waste management: Implication of technology to collect the trash, real-time tracking of the waste dump, smart bin, etc., enables the smart city to manage the waste efficiently. The connected system gathers data from several town places and routes that data to the pick-up services for waste collection. The system also monitors the methodology of dumping the waste and enables appropriate safety measurements to nullify the adverse effects.

Intelligent city engineering service empowers city administration to develop the quality of services to enhance the lifestyle of its citizens. Wise utilisation of the cutting-edge technologies results in a systematic approach towards resource management, optimisation of energy consumption and effective use of alternative energy sources.

9.6.1 Smart Energy Management

Global energy generation decreased by 0.9% in 2020, but the global COVID pandemic causes this decline. Commercial buildings and industries were closed due to global lockdown and reduced consumption of power. On the other hand, significant growth has been witnessed in case renewable energy generation in 2020. The global share of renewable energy has increased to 11.7% from 10.3% in 2019 (British Petroleum Company, 2021). But let's look back at 2019 data. It says that electricity generation grew by 1.3% carbon dioxide (CO_2) emissions increased by 0.5% globally in 2019 compared to the previous year (British Petroleum Company, 2020). The COVID pandemic is an exceptional event that causes the decline in energy demand, but the normal circumstance projects the increasing energy requirement. The positive side of this incremental demand is the growing need for renewable and alternative energy sources in the energy basket. Global thought leaders address climate change making the transition away from fossil-fuel energy generation to sustainable sources. Here the Smart Energy System (SES) plays a significant role to develop carbon-neutral growth. SES uses the intelligent solution to monitor and control the generation, storage, distribution and usage (Figure 9.10). In the SES system, wireless connectivity helps gather data from connected devices and then stored it in the cloud for analysis. The application of IoT is crucial for collecting data from multiple devices and storing it in the cloud data warehouse. AI-based technologies are helpful for cloud computation and employed for optimising usage. Finally, the platform gathers complex data from several sources and represents the respective stakeholders (GSMA, 2020).

With the expansion of urbanisation, the appetite for energy is also inflating. So, energy management has become an integral part of the city infrastructure (Lombardi, 2012). A smart city uses sustainable solutions to manage energy footprint efficiently and provides its resident with a high quality of life by incorporating Smart Energy Management (SEM). Smart energy solutions in buildings and infrastructures offer "comfort", "functionality", and "flexibility" by integrating the service value chain

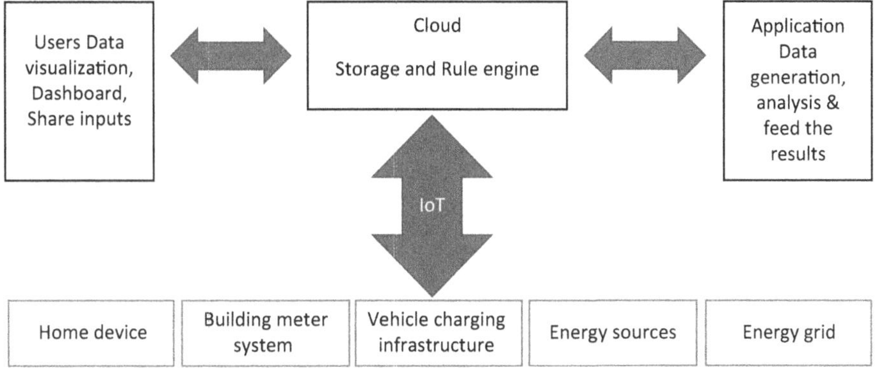

FIGURE 9.10 Flow of smart energy management system.

and implementing automation. In transportation and mobility, shifting to alternative fuel vehicles (like electric vehicles) is inspired by smart city energy management like charging infrastructure development, improving connectivity, and ensuring energy availability. Smart energy infrastructure development (like a smart grid) and data management are two critical pillars that enable and monitor complex solutions. The new generation technological solutions and tools are involved in managing, analysing, forecasting and monitoring the SEM (Mosannenzadeh, 2017).

9.6.1.1 IoT and Smart Energy Management

IoT system interconnects multiple devices and equipment and enables machine-to-machine communication. This connected network helps in accumulating quantifiable and measurable data stored in the cloud-based warehouse for analysis (Bagdadee, Zhang, & Remus, 2020). The IoT-based smart energy management system improves power quality, optimisation, online audit, and tracking consumption. Real-time health audit enables preventive measurement by capturing failure parameters.

9.6.1.2 Big Data Analytics and Smart Grids

Big data plays a crucial role in smart cities for modernising the grid system and improving operational efficiency. Data arrive from different sources (smart meters, power plants, residential homes, industries, etc.) are transmitted and pre-processed. The transmitted data are arranged in a given format with the help of a data integration technique. In the data processing stage, received smart grid data are distributed in blocks and stored in data nodes. Analytics are applied to these smart grid data to visualised and gaining knowledge about power utilisation. Learning and response are two critical objectives of analytics. The said models effectively ensure grid security, decision-making of power-sharing, load forecasting, and performance management (Munshi & Mohamed, 2017).

Smart energy management is one of the founding pillars of smart city, aims for the sustainability goal and reduction of carbon footprint. Data mining and analysis of historical records will help forecast the demand/peak load and the energy consumption

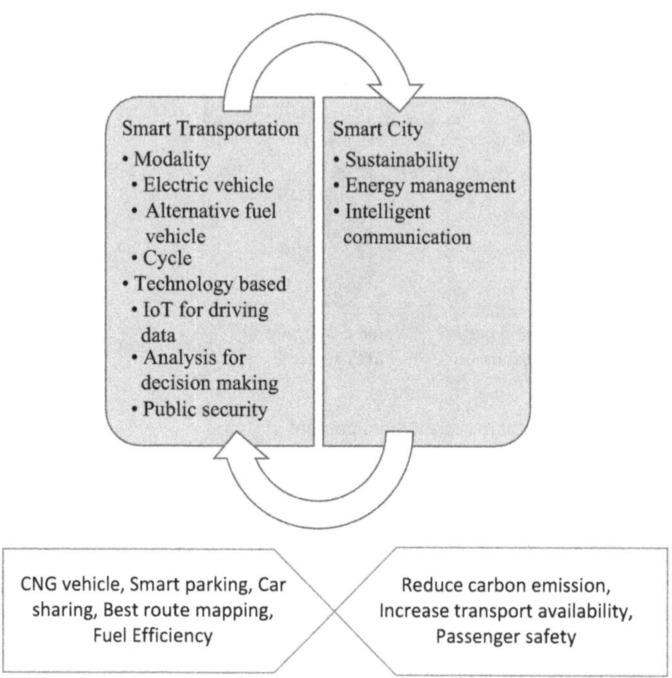

FIGURE 9.11 Relationship between smart city and smart transportation.

seasonality pattern. The said forecasting approach helps distributors optimise the energy mix in the grid (energy produced by renewable and non-renewable sources fed into the grid) and maintain power quality (Figure 9.11).

9.6.2 Smart Transportation

The fundamental objective of the smart city is to ensure the inter-relationship among human capital, social capital and infrastructure to achieve sustainability and quality of life for inhabitants. So, sustainable transport is one of the essential elements of a smart city (Bamwesigye & Hlavackova, 2019). Also, transportation is a driving factor for trade, industrial growth and economic activities. Furthermore, social life, education system, healthcare and amusement are linked with transportation by different means (Rassafi & Vaziri, 2005). Hence, a smart transportation system is crucial for managing the economic & social needs and achieving sustainability goals (Kronsell, Smidfelt, & Winslott, 2015).

9.6.2.1 IoT, Data, Analytics and Smart Transportation

IoT is the technical foundation of the smart city, equips with intelligence, interconnectivity and instrumentation. These features help in establishing communication among machines, objects and devices. The said ability is the foundation of an intelligent transportation system in the smart city. IoT-based smart transportation collect real-time data from the sensors installed in the vehicle, GPS, Smart card and roadside

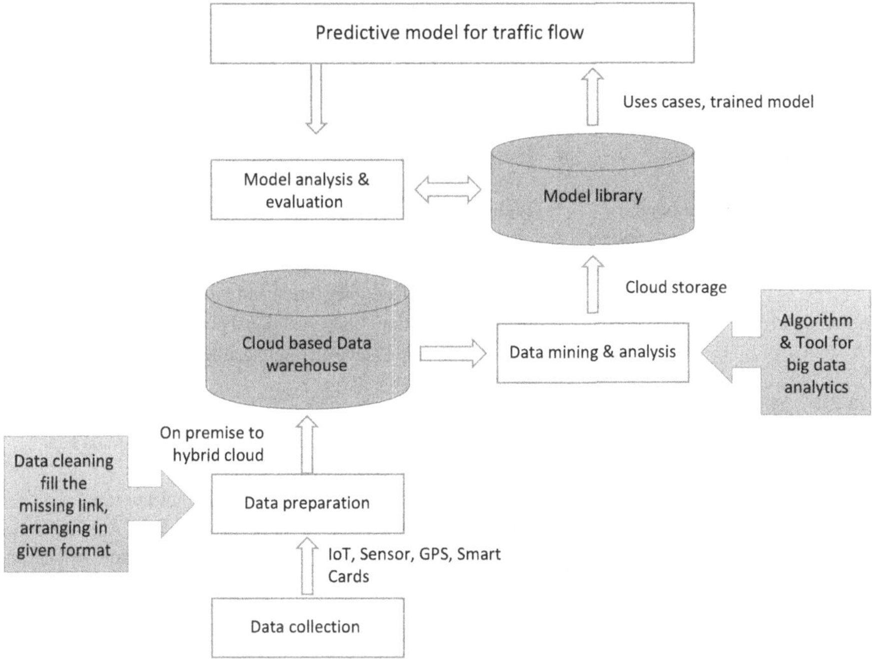

FIGURE 9.12 Traffic flow prediction model.

scanning devices, and utilises the same to provide insights to passengers like low-cost options, shortest route, automatic toll collection etc. (Figure 9.12) (Mohanty, Choppali, & Kougianos, 2016).

Moreover, the IoT-based information is stored in the cloud data warehouse. The said is fed for AI and big data analytics to develop optimised and effective decision-making tools and applications. Big data analytics is helpful for road traffic accidents prediction and prevention models by taking inputs from historical weather conditions, traffic light flow data, highway accident history, etc. Using data mining and analysis traffic flow model is established. Also, the public transport service model is developed from riders' traveling route patterns. Big data analytics empower city administration in intelligent traffic management, public transportation planning and ensuring safety. The railway transportation system is the brightest example of analytics best practices. Analysing the data Big data analytics plays an essential role by reducing traveling time, traffic congestion, and emissions of pollutants (Zhu, Yu, Wang, Ning, & Tang, 2018).

9.6.3 Smart Buildings

Technological advancement has transformed structural changes in modern urban life. Today, modern buildings are advance compared to four walls and a roof. In reality, homes are more competent enough to sense the requirement of their inhabitants. Several wireless connectivity in the buildings generate a massive amount of

TABLE 9.3
Relationship among Different Scopes and Components in Smart Building

Scope	Components	Systems
Monitoring	Safety	Fire safety and prevention, Intelligent alarming system, Communication, Remote home monitoring
	In-home environment	Light, Heating, Ventilation and Air conditioning, Energy management
	Comfort	Appliance control, Automation, Ventilation and Air conditioning, Communication and Mobility
Home Control	Satisfaction	Integrated solution, multi-functional products
	Technology	Connected Media and Entertainment devices, Voice navigation, Energy efficiency
	Flexibility	Appliance control, multi-room connectivity
Environment	Energy	Integrated sensor-based energy management, Residual heat utilisation, use of renewable energy
	Ecology	Recycling energy storage, Renewable, Rainwater use
	Efficiency	AI to analyse real-time data for decision making, Automation to manage lighting level, Temperature and Ventilation, IoT based resource use monitoring, Smart communication

data utilised to develop the quality of life and efficiency. Hence, an intelligent building can respond to real-time events with the help of machine learning automated structural models. Several sensors collect in-house ambient data like temperature, humidity, CO2 level, etc. Security devices, CCTV cameras and alarming systems are installed around the house. These connected IOT-based devices gather large chunks of user data and the analysed and processed effectively into actionable insight.

One of the core elements of the smart home management system is energy management. The critical objective of home energy management is to optimise the demand side-improve efficiency in energy consumption. Demand response management benefits users in cost management by indirectly decreasing consumption (Kabalci, Kabalci, Padmanaban, Holm-Nielsen, & Blaabjerg, 2019).

The mention points are some of the benefits derived from smart building infrastructure. The intelligent system monitors the building performance and adds value to the building productivity (Table 9.3).

9.6.4 Smart Waste Management

Smart garbage management is essential to ensure public health safety and the prevention of pollution in smart cities. Developing countries face a severe challenge related to waste management and environmental contamination due to a lack of infrastructure. The enablement of digitalisation will bring a suitable solution for the said challenge. A sensor-based efficient garbage management system measures the trash level and informs the authority when it becomes full (Sohag & Podder, 2020). IoT based waste management system empowers smart city authority to monitor the debris level over the web-based application and saves time and operational cost. The administration can initiate the collection process once the bin is filled up, and geotagging will help navigate (Nirde, Mulay, & Chaskar, 2017). The application of cloud computing will help to store real-time data gathered from sensors. Big data analytics develops the best route mapping and insights about waste generation and disposal (Aazam, St-Hilaire, Lung, & Lambadaris, 2016). The application-based waste management system provides volume status to the end-user and shares the bins where to dump. It also notifies the scheduled trip routes of the truck so that people can drop the waste accordingly (Nasar, Karlsen, & Hameed, 2020). IoT provides an affluence amount of data, and several analytics methodologies offer potential solutions to waste management with sustainability purposes.

9.6.5 Cyber Security

Smart cities intend to foster economic growth, sustainability, and quality of life by boosting convenience, adopting new technologies and services, and developing communication. Implementation of new-edge technologies like Big Data, AI, ML, IoT, Cloud computing etc., fuels the transformative changes. So, it's become a priority to monitor the security side of the digitalisation aspects of the smart city by implementing techniques and tools to ensure IT security compliance, manage the incident and enact the continuity plan.

A planned cyber-attack may gravely impact the wide range of connected devices, defence systems, energy management systems, mobility control systems, etc. Taking control over the complex city infrastructure (like Smart grid) and operation will disrupt the normal day-to-day activities ((Alibasic, Junaibi, Aung, Woon, & Omar, 2017). Some burning factors will influence cyber risks in smart cities.

- Lack of security testing: Incapability to perform vulnerability assessment and system health check-ups periodically will enhance the risk for the cyber-attack (Cerrudo, 2015)
- Poor product management: Monitoring and management are required on the operation side, and it is essential to check the stability and logging reports. Evaluation of security impact is crucial while integrating a new product in the system and deployment of measures accordingly (Cerrudo, Hasbini, & Russell, 2016)
- Convergence of IT infrastructure with the physical world makes less distinct the barrier and poor integration of services across domains through digital technologies (Pandey, Li, & Peasley, n.d.)
- Lack of developing authentication, confidentiality, intrusion detection and privacy protection weaken the infrastructure (Cui, Xie, Qu, Gao, & Yang, 2018).

Security assessment and protection technologies (Table 9.4):

> Cybersecurity analysis of the smart city infrastructure helps to identify the vulnerability, threats and risk involved. There are three types of risk assessment procedures: qualitative, quantitative and mixed.
>
> (Kalinin, Krundyshev, & Zegzhda, 2021)

TABLE 9.4
Security Assessment Tool

Assessment	Assessment Type	Description	Reference
Qualitative	Expert assessment	Subject matter experts (SMEs) will review the security architecture and help in decision making. Brainstorming, SWOT analysis, survey etc. are part of this method	Zhang and Li (2011)
	Rating	Rating and ranking processes to understand the security vulnerability	Kalinin et al. (2021)
	Checklist	Fact-checking of the risk factors. The checklist is prepared taking data from past	Kalinin et al. (2021)
Quantitative	Analytics and Probabilistic method	Sensitivity analysis, scenario testing, model creation from historical data, simulation-based models are used. New edge technology like AI-enabled security systems, use of ML	Kalinin et al. (2021) Kara and Fırat (2018)

9.6.5.1 Implementation of AI and ML for Cybersecurity

AI-enabled cybersecurity system is a booming trend in the information technology sectors. Smart cities are the hubs for the massive amount of people and administration-related data. So, it is a prime objective of the smart city administrator to conduct vulnerability assessment time-to-time and ensure security. AI can analyse enormous amounts of data and detect security threats in real-time or predict based upon modelling. AI is trained with the help of structure and unstructured data. Then machine learning and deep learning techniques improve the knowledge bank of AI about cyber risk. AI can identify the relationship between threats and respond quickly. The cognitive security model integrates human intelligence and machine power to improve the system for better risk management (IBM Cloud Education, 2020a, 2020b.). AI can bring tactical and strategic advantages in the cybersecurity domain. AI helps in detecting and minimising vulnerability.

On the other hand, strategically, AI identifies the probable threats for the defence system in advance. AI contributes towards the improvement of the system in three ways- improving "*robustness*", advancing "*response*" time and enhancing "*resilience*" (Taddeo, McCutcheon, & Floridi, 2019). AI boosts **system robustness** by identifying the undefined situation and act upon that change. Self-testing and self-healing capabilities ensure robustness and improve the system's continuous learning (Tariq et al., 2019). AI enhances robustness by upgrading security control. ML develops the predictive model for vulnerability classification, risk analysis and gap analysis. The said model is used for designing automated planning for assessment (Khan & Parkinson, 2018). The core objective of system resilience improvement is to take action against threats by adopting necessary changes. AI utilises the threat and anomalies detection (TAD) model (Taddeo, McCutcheon, & Floridi, 2019) to identify hazards and take preventive actions. Malware detection, network intrusion, monitoring phishing activities and identifying sophisticated changes in attacking techniques are used to design an AI-based preventive model that can handle potential threats to the system and enable solutions for that (Truong, Diep, & Zelinka, 2020). System resilience provides intelligence-related security information, and AI and ML-enabled security tools reduce the **response time** for the incidents (Gil, 2019).

Information and communication technologies (ICT) improve urban life in smart cities. Connected devices and systems generate a significant amount of data that contribute to enhancing city infrastructure and life. Undesired cyber-attack may damage vital services like energy distribution, traffic management, resource distribution and public facilities. Sensitive data leakage can harm citizens and city infrastructure. Cyber-security training, implementation of cutting-edge technologies, periodically maintenance and testing ensure protection wall against system attack.

9.7 CONCLUSION

The city is an integral part of the human journey. From the very beginning of human civilisation, the concept and architecture of the city have evolved with time. The quality of life, better utilisation of resources and efficient management of the city system define the prosperity of urbanisation. The implementation of digitalisation helps to centralise the systems, and analysis of the citizens' data assists in enhancing the service delivery model. IoT applications and connected devices develop a pool of user's data. AI, ML applications cultivate the said data bank to automate the system and make it intelligent. Robust application of these cutting-edge technologies improves the service delivery strategy and personalise the services, bringing delight and satisfaction. The growth of 5G technology and robotic process automation has nourished the said digital endeavour. The process of digitalisation brings some challenges. Capital expenditure (CAPEX) is a significant concern for implementing digitalisation, and the advancement carries an inherent threat of data breaches and cyber-attacks. The Government body, IT industry and corporate houses should take a deep dive to identify the suitable source for capital. The security concern can be acknowledged by adopting prescriptive and preventive measures and imposing a necessary code of conduct. The government should focus on the policy designing part while the tech-side of the smart city can develop collaborating with the leading IT companies and start-ups. The shared service model, application development and maintenance and security management by private sectors will provide scope for innovation, growth and, efficient asset management. Digitalisation is the driving force to improve the city infrastructure and nullify operational inefficiencies resulting in citizen delight and convenience.

The term "Digital transformation" is frequently used to explain the modernisation approach for legacy system, process or services, and IT centric technology (Table 9.5). If we use the term "digitalisation" as noun, it explains something need to purchase or installed. On the other hand, adjective form of "digitalisation" describes the improvement of the operational processes and enhancement of the system. There is a visible difference how the said term is used. Digitalisation does not mean doing the old work in a newer approach. Rather it should be linked with enhancement, information-intensive, data-driven approach and agility (McDonald, 2020).

TABLE 9.5
Future Scope for Smart City Digitalisation Using AI and ML

Scope	Futuristic Approach	Reference
Smart energy management	**Intelligence distribution:** Real-time demand analysis and managing gris operation using AI. Self-learning algorithm is used to administer systemic disturbances and fix the issue	Jacobson and Dickerman (2019) Espe, Potdar, and Chang (2018)
	Energy trading: Algorithm is used for demand response and management of dynamic pricing, scheduling and economic efficiency.	Antonopoulos et al. (2020) Saqib Ali and Choi (2020)
Smart transportation	**Connected vehicle:** The real time vehicle and driving data is collected through IoT and transfer to the cloud storage for further analysis. Vehicle-to-cloud service helps in road casting information to the drivers and promotes the intelligent driving practice. The said approach is helpful for navigating shortest route, locating the refuelling station (gas filling station or electric charging station), and assisting the driver.	Automotive Edge Computing Consortium (AECC) (2021)
Cyber Security	**Cognitive computing** technology has designed the system in a way that can imitate human intelligent. Analysing a vast volume of structure and unstructured data, system can identify early indicator of threat and possible impacts. Cognitive outlook-based security system plays essential role in energy management, ensuring data privacy and security and safeguard public infrastructure from external attacks	Koslowski and Felle (n.d.) Mathew (2020)

In the context of the smart city, digitalisation process plays a crucial role to modernise and enhance the public infrastructure by collecting user data that eventually add value to the public service, safety and sustainability.

Apart from the above-mentioned areas, capabilities could be built around the applicability of IoT and Blockchain. Adoption of disruptive technology like Distributed Ledger Technology (DLT) can assist government to collect tax, other administrative transactions, energy trading and social security. The said decentralised system ensures the identity and access management, data privacy, transaction and secure communication (IBM, n.d.-a, n.d.-b) (Antal, Cioara, Anghel, Antal, & Salomie, 2021). With the rapid development of 5G technology, the applicability of IoT based smart system in the green building construction is shining brighter. Leveraging the potential of digital innovations and new edge technologies is the key driving force to develop the future of the smart city and ensure well-being.

REFERENCES

Aazam, M., St-Hilaire, M., Lung, C.-H., & Lambadaris, I. (2016). Cloud-based smart waste management for smart cities. *2016 IEEE 21st International Workshop on Computer Aided Modelling and Design of Communication Links and Networks (CAMAD)*. doi:10.1109/CAMAD.2016.7790356.

Abdulqadir, H. R., & Abdulazeez, A. M. (2021). Reinforcement Learning and Modeling Techniques: A Review. *IJSAB International*, 174–189. doi:10.5281/zenodo.4542638.

Alibasic, A., Junaibi, R. A., Aung, Z., Woon, W., & Omar, M. A. (2017). Cybersecurity for Smart Cities: A Brief Review. *International Workshop on Data Analytics for Renewable Energy Integration*. doi:10.1007/978-3-319-50947-1_3.

Allam, Z., & Zaynah Dhunny, A. (2019, June). On Big Data, Artificial Intelligence and Smart Cities. *Cities*, *89*, 81. doi: 10.1016/j.cities.2019.01.032.

Alvarez, R. (2017). The Relevance of Informational Infrastructures in Future Cities. *Field Actions Science Reports* (17), 12–15. Retrieved July 03, 2021, from https://journals.openedition.org/factsreports/4389

Antal, C., Cioara, T., Anghel, I., Antal, M., & Salomie, I. (2021). Distributed Ledger Technology Review and Decentralized Applications Development Guidelines. *Future Internet*, *13*(62). doi: 10.3390/fi13030062.

Antonopoulos, I., Robu, V., Couraud, B., Kirli, D., Norbu, S., Kiprakis, A., … Wattam, S. (2020, 09). Artificial Intelligence and Machine Learning Approaches to Energy Demand-Side Response: A Systematic Review. *Renewable and Sustainable Energy Reviews*, *130*. doi: 10.1016/j.rser.2020.109899.

Antunes, M. E., Tanaka, Y., & Merritt, J. (2021). *Being Smart about Smart Cities: A Governance Roadmap for Digital Technologies*. World Economic Forum. Retrieved from https://www.weforum.org/agenda/2021/07/being-smart-about-smart-cities-a-governance-roadmap-for-digital-technologies/

Automotive Edge Computing Consortium (AECC). (2021). *Connected cars: On the edge of a breakthrough*. Mobile World Live. Retrieved September 27, 2021, from https://aecc.org/wp-content/uploads/2021/05/MWL_-_AECC_whitepaper_-_Design_v2.0.pdf

Bagdadee, A. H., Zhang, L., & Remus, M. H. (2020). A Brief Review of the IoT Based Energy Management System in the Smart Industry. In *Artificial Intelligence and Evolutionary Computations in Engineering Systems*. doi:10.1007/978-981-15-0199-9_38.

Bamwesigye, D., & Hlavackova, P. (2019). Analysis of Sustainable Transport for Smart Cities. *Sustainability*. doi:10.3390/su11072140.

British Petroleum Company. (2020). *Statistical Review of World Energy*. British Petroleum Company. Retrieved July 25, 2021, from https://www.bp.com/content/dam/bp/business-sites/en/global/corporate/pdfs/energy-economics/statistical-review/bp-stats-review-2020-full-report.pdf

British Petroleum Company. (2021). *Statistical Review of World Energy*. British Petroleum Company. Retrieved July 25, 2021, from https://www.bp.com/content/dam/bp/business-sites/en/global/corporate/pdfs/energy-economics/statistical-review/bp-stats-review-2021-full-report.pdf

Cerrudo, C.. (2015). *An Emerging US (and World) Threat: Cities Wide Open to Cyber Attacks*. IOActive Labs. Retrieved August 03, 2021, from https://ioactive.com/pdfs/IOActive_HackingCitiesPaper_CesarCerrudo.pdf

Cerrudo, C., Hasbini, M. A., & Russell, B. (2016). *Cyber Security Guidelines for Smart City Technology Adoption*. Securing Smart Cities. Retrieved 2021, from https://securingsmartcities.org/wp-content/uploads/2016/03/Guidlines_for_Safe_Smart_Cities-1.pdf

Chandra, S. (2019). City Gas Distribution: A Pillar for Developing Smart Cities. *Journal of Current Trends in Electrical Engineering*, *4*(1). Retrieved July 15, 2021, from http://mantechpublications.com/admin/index.php/JoCTiEE/article/view/1084

Colldahl, C., Frey, S., & Kelemen, J. E. (2013). *Smart Cities: Strategic Sustainable Development for an Urban World*. Karlskrona, Sweden: Blekinge Institute of Technology, School of Engineering. Retrieved July 15, 2021, from https://www.diva-portal.org/smash/get/diva2:832150/FULLTEXT01.pdf

Colmer, J. (2015). *Urbanisation, Growth, and Development: Evidence from India*. London School of Economics.

Cui, L., Xie, G., Qu, Y., Gao, L., & Yang, Y. (2018). Security and Privacy in Smart Cities: Challenges and Opportunities. *IEEE Access*. doi: 10.1109/ACCESS.2018.2853985.

Davis, K. (1965, September). The Urbanization of the Human Population. *Scientific American, 213*. Retrieved July 15, 2021, from http://hiebertglobalcenter.org/blog/wp-content/uploads/2013/04/Reading-7-Davis-The-Urbanization-of-the-Human-Population.pdf

Duin, Stefan van, & Bakhshi, Naser. (2018). *Artificial Intelligence*. Deloitte. Retrieved August 15, 2021, from https://www2.deloitte.com/content/dam/Deloitte/nl/Documents/deloitte-analytics/deloitte-nl-data-analytics-artificial-intelligence-whitepaper-eng.pdf

Economic and Social Council. (2016). *Smart Cities and Infrastructure*. Geneva: United Nations Conference on Trade and Development.

Espe, E., Potdar, V., & Chang, E. (2018). Prosumer Communities and Relationships in Smart Grids: A Literature Review, *Evolution and Future Directions. Energies, 11*(10), 25–28. doi: 10.3390/en11102528.

Estevez, E., Lopes, N., & Janowski, T.. (2016). *Smart Sustainable Cities – Reconnaissance Study*. United Nations University.

Gebre, Biniam, & Whitehouse, Mimi. (2018). *Essential insights: Artificial intelligence unleashed*. Retrieved June 20, 2021, from https://www.accenture.com/_acnmedia/PDF-86/Accenture-Essential-Insights-POV.pdf

Gil, Laurent. (2019). *AI brings speed to security*. Retrieved August 07, 2021, from O'Reilly Media: https://www.oreilly.com/content/ai-brings-speed-to-security/

GSMA. (2020). *Smart Energy Systems*. London: GSM Association. Retrieved 07 25, 2021, from https://www.gsma.com/betterfuture/wp-content/uploads/2021/02/Smart-Energy-Systems-Report.pdf

IBM. (n.d.-a). *Artificial intelligence for a smarter kind of cybersecurity*. Retrieved from https://www.ibm.com/in-en/security/artificial-intelligence?p1=Search&p4=43700052660419533&p5=e&gclid=CjwKCAjwmK6IBhBqEiwAocMc8oyD5IKh2Cx__EjcqYTNBRDuUtGG9nU6TNqooy06LgDDarXiXxuLfRoCsnkQAvD_BwE&gclsrc=aw.ds

IBM. (n.d.-b). *What is blockchain security?* (IBM, Editor) Retrieved September 27, 2021, from IBM.com: https://www.ibm.com/in-en/topics/blockchain-security

IBM Cloud Education. (2020a). *Strong AI*. IBM Cloud Learn Hub. Retrieved August 15, 2021, from https://www.ibm.com/cloud/learn/strong-ai

IBM Cloud Education. (2020b). *What is Artificial Intelligence (AI)?* IBM Cloud Learn Hub. Retrieved August 15, 2021, from https://www.ibm.com/cloud/learn/what-is-artificial-intelligence

IBM Cloud Learning. (2020). *Machine Learning*. Retrieved from IBM: https://www.ibm.com/cloud/learn/machine-learning

IMD World Competitiveness Center. (2020). *IMD-SUTD: Smart City Index.*

Innovate4Climate. (2019). *Four Things You Should Know About Climate-Smart Cities.* World Bank Group.

International Finance Corporation. (2018). *Climate Investment Opportunities in Cities.* Washington, D.C.

Jacobson, D., & Dickerman, L. (2019, 06). Distributed Intelligence: A Critical Piece of the Microgrid Puzzle. *The Electricity Journal, 32*(5), 10–13. doi:10.1016/j.tej.2019.05.001.

Jeff Merritt, M. E. (2021). *Governing Smart Cities: Policy Benchmarks for Ethical and Responsible Smart City Development*. Geneva, Switzerland: *World Economic Forum.*

Jones, M. T. (2017). *Models for machine learning*. Artificial Intelligence. Retrieved August 08, 2021, from https://developer.ibm.com/articles/cc-models-machine-learning/#reinforcement-learning

Kabalci, Y., Kabalci, E., Padmanaban, S., Holm-Nielsen, J. B., & Blaabjerg, F. (2019). Internet of Things Applications as Energy Internet in Smart Grids and Smart Environments. *Electronics*. Retrieved August 03, 2021.

Kalinin, M., Krundyshev, V., & Zegzhda, P. (2021). Cybersecurity Risk Assessment in Smart City Infrastructures. *Machines*. doi:10.3390/machines9040078.

Kara, M., & Fırat, S. (2018). Supplier Risk Assessment Based on Best-Worst Method and K-Means Clustering: A Case Study. *Sustainability*, *10*, 1–25. doi:10.3390/su10041066.

Khan, S., & Parkinson, S. (2018). Review into State of the Art of Vulnerability Assessment using Artificial Intelligence. *Springer*, 3–32. doi:10.1007/978-3-319-92624-7_1.

Khanna, S. K. (2019). Machine Learning v/s Deep Learning. *International Research Journal of Engineering and Technology*, *6*(2). Retrieved August 15, 2021, from https://www.cirjet.net/archives/V6/i2/IRJET-V6I286.pdf

Koslowski, Thomas, Felle, Martin. (n.d.). *Cognitive computing for cyber security*. Deloitte. Retrieved September 27, 2021, from https://www2.deloitte.com/ch/en/pages/risk/articles/cognitive-computing-for-cyber-security.html

Kronsell, A., Smidfelt, R. L., & Winslott, H. L. (2015). Achieving Climate Objectives in Transport Policy by Including Women and Challenging Gender Norms: The Swedish Case. *International Journal of Sustainable Transportation*, *10*, 703–711. doi:10.1080/15568318.2015.1129653.

Kubina, M., Šulyová, D., & Vodák, J. (2021). Comparison of Smart City Standards, Implementation and Cluster Models of Cities in North America and Europe. *Sustainability*, *13*. doi: 10.3390/su13063120.

Lee, J. H., Hancock, M. G., & Hu, M.-C. (2014, 10 3). Towards an Effective Framework for Building Smart Cities: Lessons from Seoul and San Francisco. *Technological Forecasting & Social Change*, 80–99. doi: 10.1016/j.techfore.2013.08.033.

Lombardi, P. G. (2012). Modelling the Smart City Performance. *Innovation: The European Journal of Social Science Research*, *25*(2), 137–149. doi:10.1080/13511610.2012.660325.

Mathew, A. (2020, 11). Smarter Cognitive and Cybersecurity with AI. *International Journal of Engineering Science Invention (IJESI)*, *9*(11), 29–33. doi: 10.35629/6734-0911022933.

McCarthy, J. (2004). *What is Artificial Intelligence?* Stanford University, Computer Science Department. Stanford University. Retrieved August 08, 2021, from https://homes.di.unimi.it/borghese/Teaching/AdvancedIntelligentSystems/Old/IntelligentSystems_2008_2009/Old/IntelligentSystems_2005_2006/Documents/Symbolic/04_McCarthy_whatisai.pdf

McDonald, M. P. (2020, September 24). *A revised working definition of Digital Transformation*. Retrieved September 26, 2021, from Gartner Blogs: https://blogs.gartner.com/mark-mcdonald/2020/08/24/a-revised-working-definition-of-digital-transformation/

Ministry of Housing and Urban Affairs, Government of India. (2015). *Smart Cities Mission*. Government of India.

Ministry of Urban Development. (2015). *Smart Cities – Mission Statement & Guidelines*.

Mohanty, S., Choppali, U., & Kougianos, E. (2016). Everything You Wanted to Know about Smart Cities: The Internet of Things is the backbone. *IEEE Consumer Electronics Magazine*, *5*, 60–70. Retrieved from http://www.smohanty.org/Publications_Journals/2016/Mohanty_IEEE-CEM_2016-July_Smart-Cities.pdf

Mosannenzadeh, F. B. (2017). Smart Energy City Development: A Story Told by Urban Planners. *Cities*, *64*, 54–65. doi:10.1016/j.cities.2017.02.001.

Muente-Kunigami, A., & Mulas, V. (2015). *Building smarter cities*. World Bank Blog.

Munshi, A., & Mohamed, Y.-R. (2017). Big Data Framework for Analytics in Smart Grids. *Electric Power Systems Research*, *151*, 369–380. doi:10.1016/j.epsr.2017.06.006.

Nasar, W., Karlsen, A. T., & Hameed, I. A. (2020). A Conceptual Model of an Iot-Based Smart and Sustainable Solid Waste Management System: A Case Study of a Norwegian Municipality. *ECMS 2020*, 34. doi:10.7148/2020-0019.

Nations., U. (2015). *Sustainable Development Goal*. New York, United States.

Nirde, K., Mulay, P. S., & Chaskar, U. M. (2017). IoT Based Solid Waste Management System for Smart City. *International Conference on Intelligent Computing and Control Systems*. doi:10.1109/iccons.2017.8250546.

Olayode, O., Tartibu, L., & Okwu, M. (2020). Application of Artificial Intelligence in Traffic Control System of Non-autonomous Vehicles at Signalized Road Intersection. *Procedia CIRP*, *91*, 194–200. doi:10.1016/j.procir.2020.02.167.

Pandey, Piyush, Li, Timothy, & Peasley, Sean. (n.d.). *The convergence of physical and digital infrastructure: Building secure and resilient smart cities and communities*. Deloitte. Retrieved September 03, 2021, from https://www2.deloitte.com/us/en/pages/risk/solutions/smart-city-cybersecurity.html

Rassafi, A., & Vaziri, M. (2005). Sustainable transport indicators: Definition and integration. *International Journal of Environmental Science and Technology*, *2*(1), 83–96. Retrieved from http://www.bioline.org.br/pdf?st05011

Rolta. (2015). *Smart & safe cities: Concept to reality: A comprehensive solution approach*. Rolta. Retrieved July 15, 2021, from http://www.rolta.com/wp-content/uploads/pdfs/resources/Rolta-Gartner-Newsletter-Smartcities.pdf

Saqib Ali, S., & Choi, B. J. (2020). State-of-the-Art Artificial Intelligence Techniques for Distributed Smart Grids: A Review. *Electronics*. doi:10.3390/electronics9061030.

Sohag, U. M., & Podder, K. A. (2020). Smart Garbage Management System for a Sustainable Urban Life: An IoT Based Application. *Internet of Things*, *11*. doi:10.1016/j.iot.2020.100255.

Taddeo, M., McCutcheon, T., & Floridi, L. (2019). Trusting Artificial Intelligence in Cybersecurity is a Double-edged Sword. *Nature Machine Intelligence*, *1*(12). doi:10.1038/s42256-019-0109-1.

Tariq, M., Santiago, D., Chin, W., Arbon, J., Adamo, D., & Shanmugam, R. (2019). AI for Testing Today and Tomorrow: Industry Perspectives. *IEEE International Conference on Artificial Intelligence Testing (AITest)*. doi:10.1109/AITest.2019.000-3.

The Royal Society. (2017, 03). *Machine Learning: The Power and Promise of Computers that Learn by Example*. Retrieved June 20, 2021, from https://royalsociety.org/~/media/policy/projects/machine-learning/publications/machine-learning-report.pdf

Truong, T. C., Diep, Q. B., & Zelinka, I. (2020). Artificial Intelligence in the Cyber Domain: Offense and Defense. *Symmetry*. doi:10.3390/sym12030410.

Ullah, Z., Al-Turjman, F., Mostarda, L., & Gagliardi, R. (2020). Applications of Artificial Intelligence and Machine Learning in Smart Cities. *Computer Communication*, *154*, 313–323. doi:10.1016/j.comcom.2020.02.069.

Washburn, Doug, & Sindhu, Usman. (2010). Helping CIOs Understand "Smart City" Initiatives. *Forrester Research*, 5–8. Retrieved from https://s3-us-west-2.amazonaws.com/itworldcanada/archive/Themes/Hubs/Brainstorm/forrester_help_cios_smart_city.pdf

World Development Report. (2016). *Publishing and Knowledge Division*. Washington, DC: World Bank Group.

Zhang, R., & Li, D. (2011). Development of risk assessment model in construction project using fuzzy expert system. *2nd IEEE International Conference on Emergency Management and Management Sciences*.

Zhu, L., Yu, F., Wang, Y., Ning, B., & Tang, T. (2018). Big Data Analytics in Intelligent Transportation Systems: A Survey. *Ieee Transactions on Intelligent Transportation Systems*. doi:10.1109/TITS.2018.2815678.

10 The Role of Artificial Intelligence for Intelligent Mobile Apps

Mohamed Yousuff, Anusha, Vijayashree, and Jayashree
Vellore Institute of Technology (VIT), Vellore, India

CONTENTS

DOI: 10.1201/b23013-10

10.1 INTRODUCTION

Artificial Intelligence (AI) is a cognition project that considers knowledge as an entity, acquires information, analyses and learns the methodologies to express knowledge, and uses these techniques to accomplish the effect of simulating human intelligence (Duan & Xu, 2012). AI is an interdisciplinary and versatile technology, one with the potential of incorporating cognition, emotion identification, machine learning, decision-making, data storage and human-computer interaction (Lu, 2019). The unique features of intelligent mobile applications such as portability, user convenience, multimodality, accessibility, interactivity, affordability and ubiquity have made them an inevitable component of life. The smartphone applications consume 88% of mobile usage time (Blair Ian, 2021). Integrating AI and mobile applications can bring intelligence and cognitive features to apps, resulting in a wide scope of applications from agriculture to health care.

AI-based mobile apps are an absolute benefaction to doctors and other clinical faculties. It helps the concerned healthcare authorities fetching updated information about the patient's current health scenario and eventually assists in decision-making. Today is the age of the world wide web, and with smartphones permeating every nook and cranny, those living in remote locations may also deserve to find quality health care. They can use smartphones apps to arrange appointments with physicians and purchase medications. The customer trends on medication can be analysed using AI-based apps to derive insights. The primary motivating factor of intelligent apps is that they periodically monitor the patient's health situations and use such valuable data to forecast critical health events (Ahad et al., 2020; Bergier et al., 2021; Tripathi et al., 2020a; Tripathi et al., 2020b).

The health care mobile applications are revolutionized by the advent of AI. StrokeSave is an AI-based android mobile app developed to forecast an oncoming cardiovascular accident (CVA) risk, assuring diagnosis in a lesser time duration. It uses sensors to collect vascular data, namely, blood oxygen, pressure and pulse rate. StrokeSave app utilizes Machine Learning (ML), Deep Learning (DL) and Computer Vision (CV) algorithms (A. Gupta, 2019; Simionescu et al., 2020). DL-based mobile application is developed to diagnose the five most infectious skin diseases. It delivers diagnostic results to the patient without touching the body, highly recommended in critical situations of disease transmission (Goceri, 2021). Melanoma is one of the most commonly affected and fatal types of skin cancer. There is an estimate of 207,390 melanoma cancer cases in the United States alone by 2021 (SkinCareFoundation, 2021). Early identification ensures that treatment and survival are highly likely. AI-based mobile automation method is necessary to distinguish between cancerous and non-cancerous skin lesions to achieve an easy prognosis (Ech-Cherif et al., 2019).

Crop production is a process of the life cycle involving other living organisms, such as fungi. Food companies and farmers can lose up to 50% if fungal pathogens are not dealt with in the early stages of crop growth (OERKE, 2006). The availability of user-friendly and accessible support instruments guarantees crop protection and treatment measures. A mobile app with cloud computing and AI-powered architectures can be a better solution for the apparent need (Picon et al., 2019). Hypoglycemia, also called low blood glucose, is one of the urgent scenarios for remote health care. It is high in diabetic patients compared with healthy people, accounting for 6 to 10% of diabetes-related deaths (Ratzki-Leewing et al., 2018). DL-based mobile application is designed to predict the onset of hypoglycemia in the body using different patient symptoms (Pagiatakis et al., 2020).

Arrhythmia is a heartbeat-related term. It is a number of heartbeat conditions in adults which are irregular. Persons with common arrhythmia are monitored and rectified spontaneously through biomedical devices such as pacemakers and implantable cardioverter-defibrillator (ICD) (Rogers et al., 2016). An AI-integrated smartphone application encourages the identification of Cardiac Implanted Electrical Devices (CIED) such as pacemakers and ICDs in emergency and critical settings (Weinreich et al., 2019). The idea of analysing cough for practical initial diagnosis of coronavirus disease-2019 (COVID-19) and the need to check its possibility is motivated by trending research facts. The AI4COVID-19 is an AI-based mobile application designed as a screening tool for COVID-19 using cough audio signals. The typical sound of a cough is affirmed by the Convolutional Neural Network (CNN) model and then feed to multiple or parallel AI-based models for classification tasks and to determine the sample belongingness to one of three results such as COVID-19 likely, COVID-19 not likely and Test unknown (Imran et al., 2020).

This chapter is organized into nine sections (including introduction), each explains the role of AI in developing an intelligent mobile application. Section 10.2 describes the Strokesave and Strokehelp mobile application, developed to forecast the onset of stroke. Section 10.3 explains the working of skin disease classification mobile apps. In Section 10.4, the Melanoma identification mobile application is discussed. Section 10.5 depicts the implementation of the plant leaf disease classification mobile app. Section 10.6 deals with the prediction of Hypoglycemia and Section 10.7 details the detection of electronics devices deployed to monitor Arrhythmia patients using a smartphone application. Section 10.8 portrays the mobile application used to detect COVID-19 from the cough sound of the user. Finally, the chapter is concluded in Section 10.9.

10.2 CARDIOVASCULAR ACCIDENT PREDICTION

CVA is a coined medical phrase for stroke. A stroke is a severe life-menacing medical circumstance that occurs due to blockage of blood supply to a region of the brain. It is a state of emergency that requires compelling immediate treatment. Early treatment for a stroke substantially increases the chances of survival or results in minor damage to patients. The survivors can undergo problems such as slurring of voice, blindness, disarray, and paralysis. Depending on the type of CVA, the risk factor of death varies. Ischaemic CVA happens because of 85% clotting in blood vessels symptoms are settled within 24 hours. Haemorrhagic CVA is the fatal type of stroke because it ruptures the arteries carrying the blood to the brain (NHLBI, 2019). Figure 10.1 depicts the statistics of strokes according to World Stroke Organization (WSO) (World Stroke Organization, 2019).

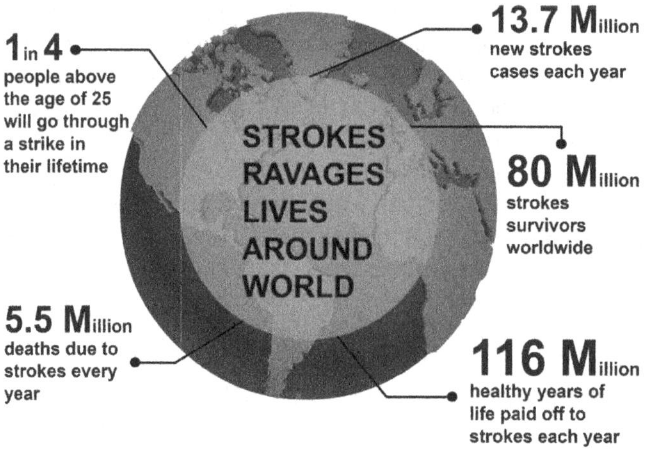

FIGURE 10.1 WSO annual report 2019.

The main CVA symptoms can be easily remembered with the acronym FAST (NHLBI, 2019) and listed as follows:

Face – Slight variations in one side of the face, drooping of mouth and eye.
Arms – Difficulty in lifting the arms; numbness is noticed in one arm.
Speech – Garbling of words during speech and confusion in understanding.
Time – Instant response to symptoms can minimize the damage.

10.2.1 StrokeSave

StrokeSave is an AI-based android mobile application meant to predict the risk of oncoming CVA, ensuring treatment in a minimum timeframe. It encompasses sensors to collect vascular data, namely, blood oxygen, pressure and pulse rate. StrokeSave app also uses ML, DL and CV algorithms to efficiently analyse and diagnose symptoms of strokes such as inconsistent heartbeats, high blood pressure, oxygen deficiency (hypoxia), hypertensive retinopathy, Bell's palsy (facial paralysis) and dysarthria (vocal paralysis) (A. Gupta, 2019).

10.2.2 Working of StrokeSave

Vocal paralysis identification module is to recognize the CVA-induced vocal palsy in the patient before the actual stroke. TORGO dataset, containing 150 dysarthric articulation audio samples, is utilized for training and testing the hybrid CNN and Recurrent Neural Network (RNN) model. The voice of the patient is recorded using the user interface of the StrokeSave app and passed through a low pass filter to minimize background noise. The filtered audio signal is then sent to the trained model via the Firebase database for learning the features and performing classification task to determine normal or palsy voice signal (A. Gupta, 2019).

Vascular data analysis and detection module is to prognosticate stroke from the sudden changes in the body, such as variation in heart pulse rate, blood pressure and oxygen levels. These data are collected through sensors present in the user body and sent to the mobile application via Bluetooth. Support vector machine (SVM) model is trained and validated on NCBI and NIH dataset of 700 vascular observations. Further, the received vascular data is then subjected to a trained SVM model for prediction. Hypertensive retinopathy diagnosis module is to detect the changes in the patient retina through an eye image taken via a lens. A deep neural network (DNN) is trained and tested using 200 images belonging to NIH eyeGENE dataset. The model then helps to determine between normal or affected retina (A. Gupta, 2019).

Facial paralysis identification module is meant to identify the facial palsy situation in the user's face. The user takes a picture of the face and sends it to the Firebase database. The CV algorithm module retrieves the face image and applies the Active Appearance Model (AAM) algorithm to divide the face into eight regions and locate 68 facial points using an annotation database. Further AAM algorithm computes dynamic displacements features of facial organs and sends it to SVM and linear regression model for facial palsy classification (A. Gupta, 2019).

Finally, the StrokeSave mobile app gives the confident percentage of risk the user has for stroke. An SVM model is trained and implemented to perform this classification task using various confident percentage levels consolidated from audio, vascular data, retina image and facial palsy. Various metrics such as F-beta score, Precision, Sensitivity, and Accuracy are measured to evaluate the performance. The StrokeSave AI-based mobile application is found to have 95% accuracy in predicting CVA or stroke (A. Gupta, 2019).

10.2.3 StrokeHelp

StrokeHelp is also an AI-based mobile application meant to predict strokes and categorizes its seriousness using the Japan Urgent Stroke Triage (JUST) score and finally notifies the current health state, location and time tokens of the patient to the emergency contacts via automatic Short Messaging Service (SMS). The architecture of the StrokeHelp app is based on FAST stroke symptoms. The app is initially calibrated with user activities such as hand movements, speech and normal facial features. The demographic data, smoking and drinking status and previous medical complication (if any) are also stored (Simionescu et al., 2020).

10.2.4 Working of StrokeHelp

Facial Module involves with observing the facial features. The user opens up the front-facing camera of the mobile and gets instructions from the StrokeHelp app's user interface to smile. The video frames of smiling and state of transition of the face while smiling is recorded as a video and given as input to the facial module. This module is designed to extract facial features from the video feed frames using an ANN-based ultramodern library called CameraX API (*CameraX API*, 2021). The facial landmarks such as nose tip, lip corners, eyeball curves are identified and made smoother to remove outliers through displacing the average of a sliding window.

In order to detect facial paralysis, the symmetric score of the mouth, eyes are computed using the geometrical model (Simionescu et al., 2020).

Arm module consists of a strength test and dexterity test. The patient suffering from an onset CVA may have difficulties in lifting the forelimbs above the head. Gyroscope sensor available in the latest mobile phones is utilized to measure the strength of hands. The dexterity test is conducted by prompting the user to connect a pattern of the same difficulty using both hands. Failing these tests confirms that the user has some sort of difficulty in forelimbs (Simionescu et al., 2020).

Speech module checks the audio signals of the user to detect aphasia (state of difficulty in speaking and understanding a language). The app prompts the user to pronouns a set of sentences and records the speech signal of the user. An integrated android speech detection module powered by AI is used to analyse the patient voice and computes the confidence score. The difference between the current, confident score and initial confident score (confident score computed during calibration phase) gives the level of slurring in the speech (Simionescu et al., 2020).

Time module plays a vital role in immediate communication to the concerned person for further treatment. Based on the data amassed using the above modules and the stored data, the JUST score is computed. If the JUST score is small or very close to 0, then the user is labelled as a low-risk patient. If the JUST score is equal to 1, then the user is a medium-risk patient. If the JUST score exceeds one, then the user is a high-risk patient. All this information is composed as a Short Message Service (SMS) along with a Google Map location link, time of symptoms and sent to the persons in the emergency contact list. A notification with the above information is popped up, breaching the screen lock security features of the mobile phone, to inform the health workers about the current state of the patient (Simionescu et al., 2020).

10.3 SKIN DISEASE DIAGNOSIS

A DL-based mobile application is proposed to diagnose the five most commonly occurring skin diseases: hemangioma, seborrheic dermatitis, psoriasis, acne vulgaris and rosacea. The strong motivation behind this app is that it provides diagnosis results to the patient without a physical touch, which is very much recommended in cases of disease transmission or a patient already infected by some other contagious illness. Consultation with a dermatologist is not always possible because of many factors such as physical inabilities, agedness, mental issues, weather conditions, job or business commitments, travelling, shortage of dermatologists, especially in villages and highly populated areas. These problems can be avoided by using automated methodologies, specifically in a handy mobile application. AI-based methods diagnose the sickness with high accuracy and promote on-time treatment with less cost (Goceri, 2021).

10.3.1 Classification Model

The mobile application uses a modified version of MobileNet architecture with a novel hybrid activation function to better classify diseases with high-performance metric. Rectifier Linear Unit (ReLU) (Goodfellow et al., 2016) activation function is commonly used in DNN architectures for the classification task, but it suffers from permanent deactivation of neurons because the weights are not updated in regions with zero gradients. Similarly, LeakyReLU activation function values get to bounce around while computing gradient descent. Drawbacks of both activation functions are overcome by using hybrid and novel loss function. The model is trained and validated on public skin lesion datasets, namely Dermweb, DermNet, Dermatoweb and DermQuest (Goceri, 2021).

The mobile application provides many options on its user interface, as shown in Figure 10.2. In the beginning, the user is requisitioned to fill up all the necessary demographic, location and personal information after a sign-up procedure. The patient is then asked to answer a few questions specifically on the symptoms, for example, how far the patient is suffering from this disease? Previous medicines? Fever or pain in the region? Now the screen is ready to receive a camera click of a picture of infected skin. Options are displayed to crop the irrelevant regions and confined with a diseased lesion on the skin (Goceri, 2021).

The cropped image is then subjected to a modified MobileNet CNN (A. G. Howard et al., 2017) network for classification and diagnosis. If the symptoms-based conclusion matches with the AI-based conclusion, then results, i.e., the skin disease with a probability value, is displayed or else the user is advised to consult in person with the dermatologist. If the computed probability value is below 85%, then also dermatologist consultation is mandatory. The app also provides information about the disease and tries to educate the user about skin illnesses via quiz sessions (Goceri, 2021).

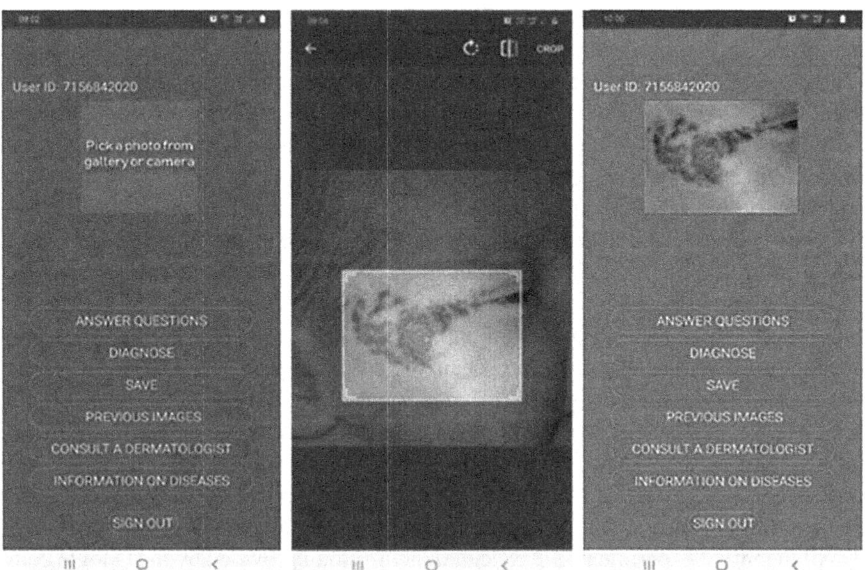

FIGURE 10.2 The main screen of the mobile application with options and demonstration of cropped lesion of skin (Goceri, 2021).

10.4 MELANOMA PREDICTION

Melanoma is a subtype of skin cancer which is one among the most widely affected and deadly cancer. Melanin secreting cells of the skin called melanocytes are the host for Melanoma. In the United States alone, there is an estimation of 207,390 melanoma cancer cases by 2021, out of which 7,180 patients are estimated to die by the end of the year. In addition, there is an annual hike of 44% in new melanoma infected cases from a period between 2011 and 2021 (SkinCareFoundation, 2021). The expert dermatologist inspects the infected skin region physically and visually and confirmation depends on medical tests and procedures, namely histopathology test report, skin lesion biopsy, and dermoscopic interpretation. Early identification of Melanoma assures a high probability of better treatment and survival. Teledermoscopy is one of the emerging techniques to consult and treat skin diseases (Horsham et al., 2016). In this technique, the patient is meant to crop and share the skin lesion images with their consulting specialists, so it involves a skilled patient activity in the diagnosis procedures. Furthermore, to achieve a convenient prognosis, a mobile automation methodology is the need of the hour to discriminate between cancerous and noncancerous skin lesions.

10.4.1 PREDICTION MODEL DEVELOPMENT

The success of DL in the classification of images with high accuracy in various domains gives hope for AI-based mobile application to predict Melanoma skin wounds as early as possible. CNN architectures are state-of-the-art in medical image classification problems. However, due to its numerous layers and trainable parameters, it is almost impossible to implement and run it on currently available mobile platforms. So, a minimized version of CNN, i.e., MobileNetV2 (Sandler et al., 2018), is used to build the model that does not require Graphics Processing Units (GPUs) to train because it constitutes only 88 layers and 3,538,984 parameters. Model training is done using standard skin lesion datasets; Dermofit Image Library, DermNet and ISIC Archive (Ech-Cherif et al., 2019).

Dermoscopy and histological images present in the datasets are well standardized since they are collected from microscope and biopsy. The photographic images originate from smartphones; thus, they suffer from various factors such as lighting, angle of orientation, zoom and focus, which affects the accuracy parameter of the classification task. The effects of photographic images can be suppressed using data augmentation techniques like mirroring, changing color, rotation and arbitrary cropping. The images in the datasets are of different dimensions. MobileNetV2 has an input size of 224 × 224 pixels, so; interpolation algorithms are employed to transform the images to fit correctly, assuring the same size as the intended network. A faster convergence rate of gradient descent and selecting an appropriate learning rate can speed up the training process, which is achieved by the normalization of the inputs.

All the initial experiments are executed on the cloud provided by the Google computation engine. Optimization of cost function and faster training of the model is achieved using *keras.utils.multi_gpu_model* data parallelism package. 4 CPUs and 4 NVIDIA Tesla V100 GPUs with 30GB of memory space are utilized for these

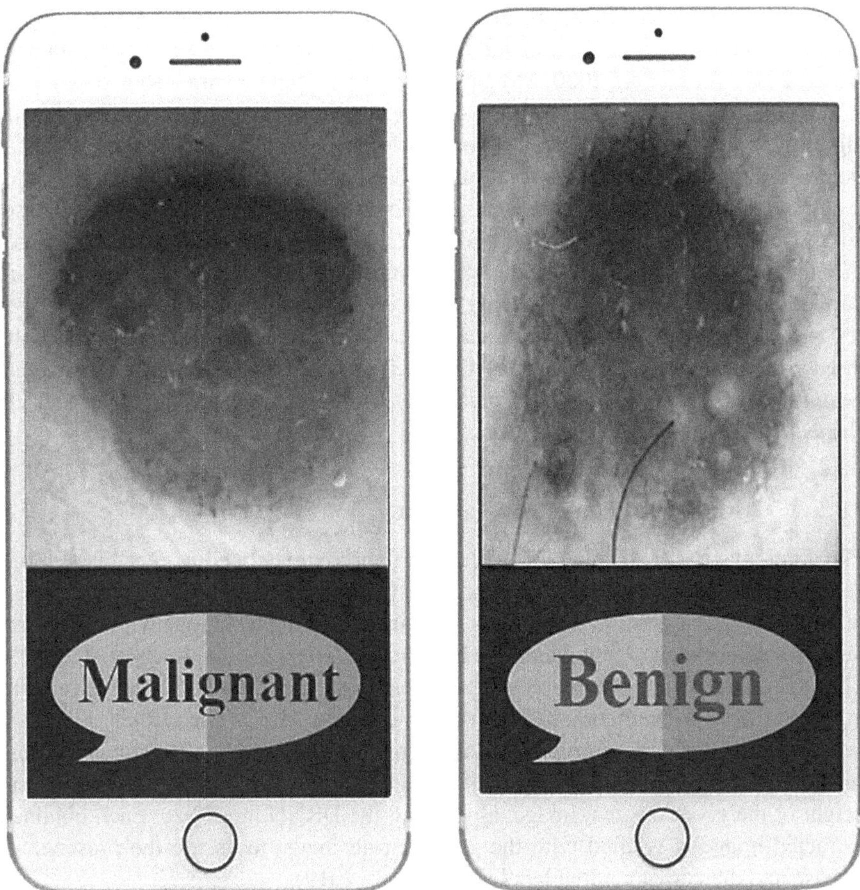

FIGURE 10.3 Working prototype of Melanoma detection iOS mobile app.

experiments. MobileNetV2 model with a batch size of 32 and Adam as an optimizer resulted in an accuracy of 91.33%. Core ML framework is used to deploy the working model as a mobile app in the iOS platform. The Figure 10.3 shows the working prototype on an iOS-based mobile phone.

10.5 PLANT LEAF DISEASE DETECTION

Food production is most important sector which directly involved and promotes life. Crop cultivation stage to harvesting stage is a life cycle process which includes other living organisms like fungus. The food production corporations and farmer can suffer a loss of up to 50% if the fungal pathogens are not addressed in the early stages of crop cultivation (OERKE, 2006). Deep knowledge of causes, lifecycle and treatment of these fungal species infestation is the key factor in saving the crop. Availability of experts is very difficult especially in huge cultivation regions. Crop protection measures and treatment is ensured with availability of user friendly and accessible supporting tools. A mobile application with a support of cloud computations and AI powered architectures can be a better solution for the undeniable necessity (Singh & Misra, 2017; Torai et al., 2020; Mohamed Yousuff & Rajasekhara Babu, 2020).

An auxiliary mobile application is developed with extensive support from DNN algorithms running on cloud, can deliver early disease identification and classification of the infecting species. A total of 8,178 images belonging to three most commonly infecting fungal species are gathered for this model development. The dataset comprises of 3338 images belonging to Rust, 2744 images of Septoria, 1568 images of Tan spot and 1116 images of healthy plant leaves. All the acquired images are perfectly segmented and labeled using a of experts. Many CNN based architectures resize the input images before classification, as a result elusive information about the diseased margins on the leaves may be lost, to avoid this problem Deep Residual neural networks (DRNN) with full picture resize, leaf mask crop and superpixel tile shaped extraction is used (Picon et al., 2019).

10.5.1 Model Parameters and Training

The input image is segmented into a cluster of approximately homogeneous sections called as superpixels. Simple Linear Iterative Clustering (SLIC) is the reliable superpixels formation algorithm which uses k-means clustering technique with configuration parameters such as compactness (200.0) and sigma (1.0) (Achanta et al., 2012). The superpixels which are not crossing the leaf mask are removed while the remaining superpixels are autonomously resized to the input size of the DRNN. Gaussian kernel is employed as a preprocessing technique to smoothen the image before segmentation process. Total number of segments can be computed using width and height of the given original image, along with the DRNN input size. Each obtained extracted image is verified with the ground truth image to ensure the presence of disease and the corresponding label (Picon et al., 2019).

The classification model is a modified version of ResNet-50 (He et al., 2016) with input picture size of 224 × 224 and 50 layers. Batch normalization and ReLU activation is used after every convolution layer. Skip connections are used to ensure establishment of deeper networks. Both in the single-branch and regular networks, degradation issue is avoided because of the ability of deeper networks to perceive complicated mappings. Two 3 X 3 convolution layers are placed in a sequence, both ensued by 3 × 3 maximum pooling layers, helps to extract fine and subtle visual features of the diseases on leaves. The actual softmax layer of 1056 output size present in ResNet-50 architecture is replaced by an output dense layer of size as same as the disease categories. Sigmoid function is used to identify many diseases on a single leaf (Picon et al., 2019).

10.5.2 Model Performance Metrics

The training process of the network is done using optimization algorithm called Stochastic Gradient Descent (SGD). The hyperparameters such as learning rate, learning rate decay and momentum are assigned with values of $10^{-4}10^{-4}$, 10^{-6} and 0.9, respectively. Before training all the layers of the complete network, the weights

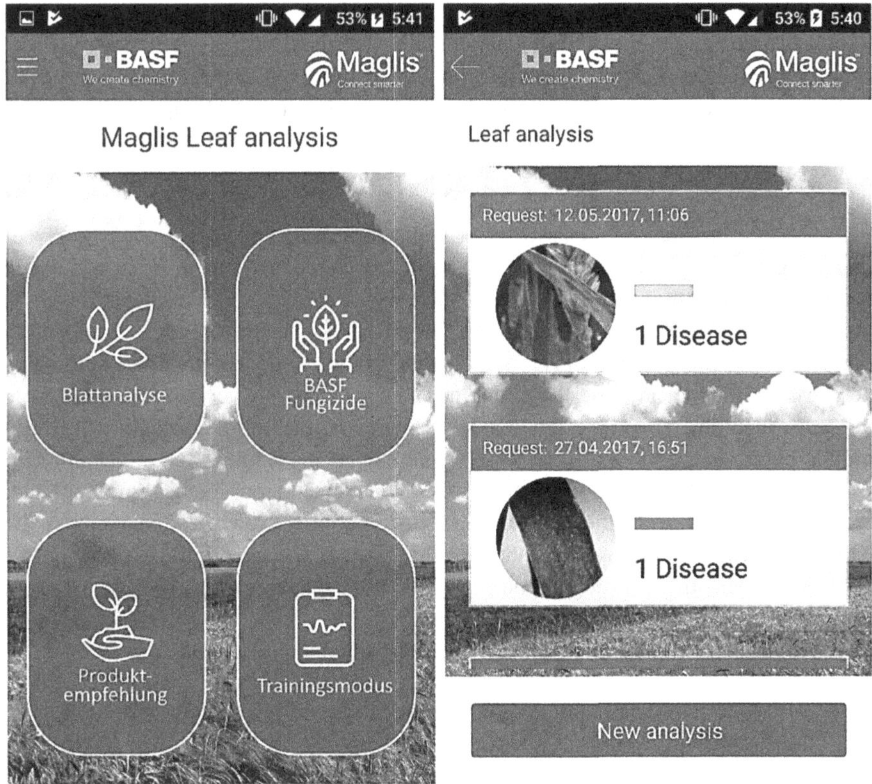

FIGURE 10.4 AI-based plant leaf disease detection mobile application (Picon et al., 2019).

of the terminal layer are first updated in 100 epochs. Bias and Variance tradeoff is entertained by splitting the dataset into three sections; 80% of the images for training, 10% for validation and 10% for testing. Balanced Accuracy (BAC) is the performance metric which is computed as an average of specificity and sensitivity. BAC is found to increase from 78% to a level of 84% in this model compare to the existing model. Well-trained, validated and tested plant leaf disease detection model is deployed in the smart phone as an application to check the real-time performance. The application gives support to get installed in Windows, Android and iOS mobile platform. Figure 10.4 depicts the mobile application (Picon et al., 2019).

10.6 HYPOGLYCEMIA MANAGEMENT

The patient-focused individualized health care domain has become more eminent, improvised and cost-effective using remote health services combined with health information techniques. Statistical data between 2018 and 2023 conveys a valuable insight that the remote patient observation and controlling (medical devices) market is expected to rise at an annual growth rate of 27.5%, attaining a global market value of above $29.9 billion (B. R. Staff, 2019). Many medical circumstances are addressed using these remote monitoring frameworks by providing services like disease awareness, education and training for the patient to handle the health crisis during the onset of an unusual health event. However, the fact of matter is that these remote health tracking platforms lack automatic analysis and a self-care-driven approach in deteriorating patient health conditions. Implementing AI-based techniques can guarantee analysis and prognostics with high efficacy, hence assures more effective service during emergencies (Pagiatakis et al., 2020).

One concerning urgent scenario plying well to remote health self-care is hypoglycemia, also termed low blood glucose. It has a high frequency of occurrences in diabetic patients than healthy individuals and even accounts for 6% to 10% of diabetes-associated deaths (Ratzki-Leewing et al., 2018). DL-based mobile application is developed to predict the oncoming of hypoglycemia condition in the body using various symptoms perceived by the patient; thus, the problem can be avoided, managed, or treated on time to escape fatalities. Therefore, the goal is to devise and incorporate a smart interaction mobile phone-based model for hypoglycemia management into a remote health care framework. The application should first adopt the conventional medical protocol and, secondly, adjusts in real-time. The driving elements for the dynamic adjustment are the patient's fugacious cognitive and physical disabilities. The intended AI-based mobile application constitutes of three modules as follows (Pagiatakis et al., 2020):

- Event detection, or in other words, detects the oncoming or the viable onset of an incident.
- Symptom assessment, or in other words, the analysis of the user's physical and cognitive condition.
- Protocol-based interference delivery based on patient health status.

10.6.1 Event Detection

Patient observation with the help of consolidated patient data is the key factor behind the event detection module. In this event identification component, the data is integrated, especially from the patient's demographic information, Electronic Medical Record (EMR), data accumulated from medical-grade sensors, and manually entered patient data. Data analysis and DL techniques are used to predict the oncoming of an emergency. Detecting the onset of hypoglycemic incidents starts by measuring the blood glucose level of the patient (Pagiatakis et al., 2020). The threshold value of 4.5 mmol/L is considered to indicate the risk of a hypoglycemic event. The level above 4.5 mmol/L is normal, glucose value between 4.0 mmol/L and 4.5 mmol/L is mild, the reading up to 2.8 mmol/L is critical and the glucose value below 2.8 mmol/L is the emergency condition necessitating external help (Yale et al., 2018). The AI-based mobile application uses physiological data in case of a mild situation to predict the event. The cognitive assessment module is invoked to predict the incident in case of a critical situation and the protocol-oriented intervention deliverance module is invoked in case of an emergency (Pagiatakis et al., 2020).

React Native mobile application development framework is used to develop the user interface of the application. Dexcom, Inc provides the data required to train and test a model. The dataset contains glucose readings of a patient for 10 hours, containing values for all the possible states (normal, mild, critical and emergency) of oncoming hypoglycemic events. The glucose reading is measured using Dexcom consistent glucose monitor with a frequency-time of 5 minutes. The onset of the hypoglycemic case also results in a feel of palpitation and anxiousness in the patient. So, physiological data such as heart rate and galvanic skin response (GSR) are also utilized in prognostication. GSR is the measure of skin resistance to a particular stimulus and an indicator of stress. The aforementioned physiological data is provided by Shimmer sensing company (Pagiatakis et al., 2020).

10.6.2 Symptom Assessment

The symptom evaluation module is invoked with a condition of mild (up to 2.8 mmol/L) glucose level reading. Vision test, Speech test and Dexterity test are performed using the mobile application to evaluate the onset of hypoglycemic episodes. The facial features and monitoring are the inputs to the vision test. The patient is requested to focus the face in the front-facing camera present in the smartphone. DL-based Google mobile service library is accessed through a react-native package to measure the capability of the patient to align the face in the camera view and compute left and right eye-opening probability. The eye-opening threshold probability (0.6) is compared with patient's eye-opening probability to detect compromised vision (Pagiatakis et al., 2020).

The speech test component gets the input from the microphone, usually present on the lower side of the smartphone. Mini-Mental State Examination (MMSE) is a benchmark test assessing the cognitive capabilities of patients. A sentence (The cat always hides under the couch) approved by MMSE is asked to be pronounced by the patient (Arevalo-Rodriguez et al., 2015). The sentence uttered by the patient (transcribed sentence) is analysed by a react-native voice package. The difference between original and transcribed sentences is measured and compromised speech impairment is thus identified. The dexterity test is performed using a mobile integrated accelerometer. The patient is kindly instructed to hold the mobile for a stipulated time. The tremors during the course of tests cause fluctuation in accelerometer reading. React native sensors package is used to analyse the variations in accelerometer reading to notice compromised dexterity (Pagiatakis et al., 2020).

10.6.3 Protocol-Oriented Intervention

This module is accessed in the final stage, i.e., emergency condition (glucose level < 2.8 mmol/L). The patient is asked to imbibe a dose (15 mg) of glucose and then take manual glucose level reading after 15 minutes. The same procedure is requested to attempt a second time if the glucose level is below 2.8 mmol/L. External interference is mandatory and highly recommended on consistently low levels of glucose in the blood. The Figure 10.5 shows the hypoglycemia management mobile application (Pagiatakis et al., 2020).

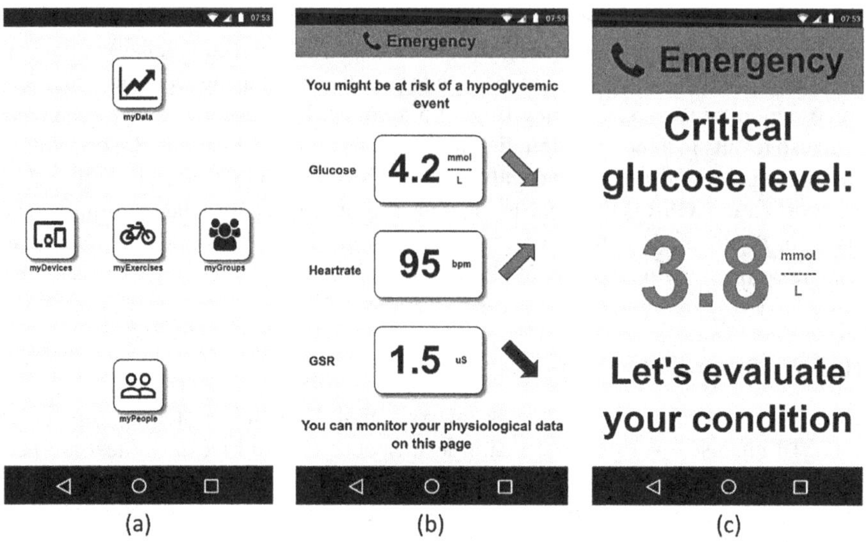

FIGURE 10.5 Screenshots of AI-based hypoglycemia management mobile application (Pagiatakis et al., 2020).

10.7 CARDIAC IMPLANTED ELECTRICAL DEVICES DETECTION

Arrhythmia is a term associated with the heartbeat. It is a set of irregular states of heartbeat in adults, pretty fast, i.e., beyond 100 beats/minute is known as tachycardia and awfully slow, lesser than 60 beats/minute is called bradycardia. People with periodical arrhythmias are incessantly monitored and spontaneously rectified using biomedical devices like pacemakers and ICD. A pacemaker is a small device that functions on battery power, deployed into the patient body to improve slow heart pumping rate. ICD is a comparatively bigger device, installed under the collarbone a shown in Figure 10.6, to manage both slow and fast heartbeat rates (Rogers et al., 2016).

A smartphone application possibly promotes the identification of CIEDs such as pacemakers and ICD in emergency and critical settings. Configuration and interrogation of CIEDs involve proprietary software and delicately crafted branded equipment. During an emergency, the identification card of the device and other appropriate details may not be promptly available. Currently, identification of the device depends upon the manual analysis of chest radiographs. Hence, it is complicated and time-taking (Jacob et al., 2011). Recently, the CNN model has been demonstrated to determine CIED on chest radiography (J. P. Howard et al., 2019). A similar approach is implemented to develop a user-friendly mobile phone-based health care application for forefront clinicians to apply such AI algorithms. The development of medical AI methodologies is essential, and implementing these progressions to an easily accessible smartphone is worth challenging.

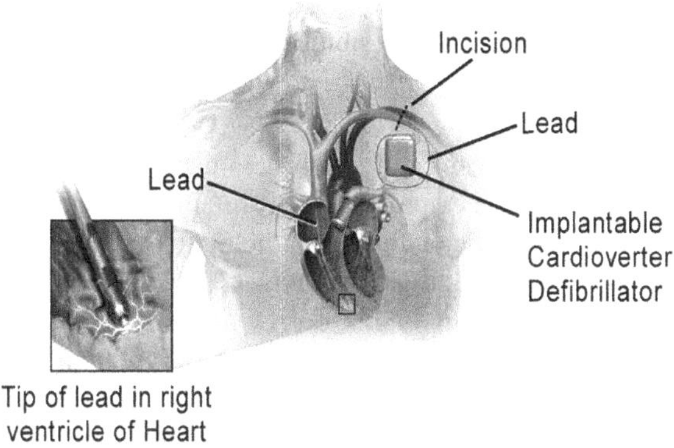

FIGURE 10.6 An ICD is placed beneath the skin of the collarbone to detect arrhythmias and reacts with electrical signals to renovate the heart's normal rhythm (B. co. Staff, 2014).

10.7.1 Model Development and Training

The anteroposterior and posteroanterior chest radiographs from CIED implanted patients over three years from 2016 to 2018 are obtained. The collected images are categorized based on device manufacturers like Abbott, Biotronik, Medtronic and Boston Scientific. Raw X-rays are resized to 400 × 400 pel red, green, blue (RGB) singe-channel files. Various data augmentation techniques, including random cropping, contrast variation, brightness adjustment, vertical and horizontal flipping, are performed. Mobile phone snapshots are also included to integrate artifact deviations. The images captured through mobile phones are subjected to diversified ambient lighting effects. Tensorflow (Martín et al., 2015) and Keras (Chollet, 2015) libraries of Python are utilized for the development of the model. K-fold cross-validation with all unique datapoints in a ratio of 7:2:1 is taken for training, validation and testing the model. Performance of image classification task is evaluated using was Tensorflow analysis modules. Sensitivity and specificity values are computed from confusion matrix parameters. Statistical significance is assessed using Pearson's chi-squared test (X2) (Weinreich et al., 2019).

A total of 3,008 images constitutes both unique and augmented radiographs. Almost 53% of the radiographs belongs to ICD and 47% corresponds to pacemaker deployments. According to the ratio of 7:2:1 2,106 images are used for training the model while 602 images for validation and 300 images for testing purpose. The installed AI-based mobile application model achieves a classification accuracy of 100% for Abbott images, 91% for Biotronik images, 94% for Medtronic images and 95% for Boston images, respectively. The receiver-operating characteristic curves metric was found to be > 0.95. The k-fold cross-validation was performed on a model with seven intermediate or hidden layers consisting of 8, 16, 32, 64, 64, 64 and 128 neurons, appropriately. The model resulted in a loss of 0.11 after 23 epochs. The overall validation accuracy is marked as 97% with 98% specificity, 95% sensitivity and X2 yielded a value of $p < 0.001$ (Weinreich et al., 2019).

10.8 COUGH SOUND SCREENING FOR COVID-19

The notion of analysing cough for feasible initial diagnosis of COVID-19 and the demand to inspect its possibility is inspired by the following salient research facts.

- Cough from trenchant respiratory disorders has distinct potential features (Miranda et al., 2019). The application of suitable mathematical techniques and signal processing on cough sounds helps to extract these unique features. Since COVID-19 uniquely damages the respiratory system, there are variations in cough sounds of COVID-19 patients and non-COVID-19 patients. Training an intricate AI model with these features results in discrimination of cough sounds.
- In 67.7 % of COVID-19 patients, cough is an apparent symptom but not in all infected patients, still coughing is one of the leading causes of social dissemination of COVID-19 (Cohen, 2020; Organization & others, 2020).
- The virus present in the cough droplets sticking on surfaces where the organism has been shown to live for more extended periods has been stated as the most severe means of spreading the COVID-19 (van Doremalen et al., 2020). Hence, the COVID-19 patient with cough as a symptom is probably spreading more infections than the non-COVID-19 counterparts.
- Temperature scanning is the most widely used screening technique for COVID-19. The number of non-COVID-19 medical conditions that can cause fever is substantially more than non-COVID-19 diseases causing cough. Hence, cough can also be considered as a pre-screening approach by requesting the user to simulate cough.

10.8.1 Working of AI4COVID-19 App

The AI4COVID-19 is an AI-based mobile application that is developed as a screening tool for COVID-19 using cough sounds. The user interface of the mobile application collects the cough sample from the user. The recorded cough sound is further verified for surrounding noise, discrepancies and corrupted cough. The typical sound of a cough is assured by the CNN model and then subjected to multiple or parallel AI-based models for classification tasks and to determine the sample belongingness to one of three outcomes such as COVID-19 liable, COVID-19 not liable and Test ambiguous. The cough sound dataset constitutes samples of COVID-19, bronchitis and pertussis. The COVID-19 data includes both spontaneous (symptomatic) cough and non-spontaneous (prompted to) cough samples. The working of AI4COVID-19 mobile application is shown in Figure 10.7.

10.8.2 Cough Detection

A cough detector module is placed before the cough discrimination module to check whether the recorded user sound is a cough or not a cough (because of background noise overlapping condition). Since AI4COVID-19 is designed to get deployed in public places (for example, shopping malls, railway stations, bus terminus and

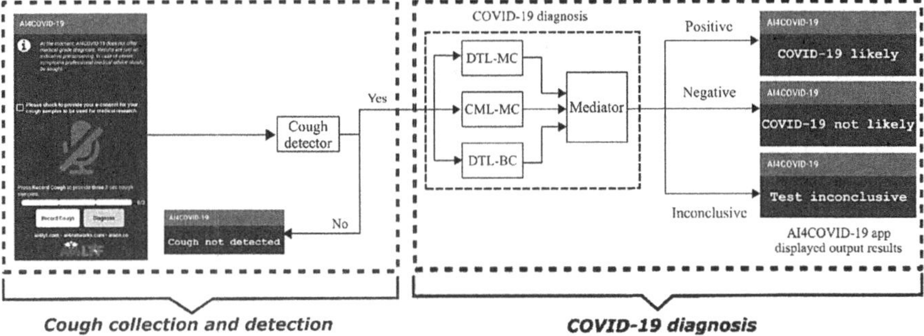

FIGURE 10.7 AI4COVID-19 mobile application collects the cough sample of the user from the input interfaces and forwards the valid cough sound sample to the AI-based analysis module for COVID-19 detection (Imran et al., 2020).

airports), the user cough sound may be perturbed with background noise. To overcome these issues, a CNN model is implemented to distinguish the sample in such a way that, in case of more noise, the user is asked to re-record the cough sound, and in case of quality cough input, the cough detector module forwards the cough sample for classification. CNN model is trained on 1839 cough sounds and 3598 environmental sounds (non-cough) belonging to 50 classes (Piczak, 2015).

10.8.3 Cough Sound Classification and Metrics

Mel-spectrogram with 128 bands is computed from the input cough sounds. The Mel-spectrogram images are grey scaled and resized to 320 × 240 × 1 dimension and fed as input to the CNN model to classify cough and non-cough sounds. The model complexity is reduced before forwarding to the next layers using a 2 × 2 max-pooling layer. The first block contains two convolutional layers with a kernel size of 5 × 5 and 16 filters. Similarly, the second block contains two convolutional layers with the same kernel size but 32 filters. Each previous block is preceded by a 2 × 2 max-pooling layer and a dropout (0.15) layer. The complicated features which are extracted are then flattened out and passed to 256 neurons arranged in a fully connected manner. A dropout (0.30) layer is placed next to avoid overfitting. At last, two neurons output layer with softmax activation is placed to classify the cough sounds. All the convolutional layers are implemented using the ReLU activation function. Backpropagation algorithm with Adam optimizer is used for updating the weights and binary cross-entropy is used as the cost function (Imran et al., 2020).

The COVID-19 diagnosis module incorporates three classification models, namely, Deep Transfer Learning-based Binary Class classifier (DTL-BC), Deep Transfer Learning-based multi-Class classifier (DTL-MC) and Classical Machine Learning-based multi-Class classifier (CML-MC) (Pan & Yang, 2010). The mobile application renders a diagnosis as "COVID-19 liable" or "COVID-19 not liable" only if all three classification models conclude the same classification outcomes. On the disagreement of the outcome by the classifiers, the mobile applications display "Test ambiguous" as a result. The models are trained and tested using 131 pertussis,

97 bronchitis, 248 normal and 72 COVID-19 cough sounds collected from various people. The objective of this tri-model architecture is to reduce the probability of misclassification or misdiagnosis (Imran et al., 2020).

DTL-BC uses a transfer learning approach along with CNN. The Mel-spectrogram images of all the cough samples are subjected to the CNN model with the same architecture as implemented in the cough detection module. It is a binary classifier that concludes whether the given cough samples are associated with COVID-19 or not. DTL-MC also uses transfer learning and CNN architecture as used in the case of DTL-BC but with a slight difference in the output layer. Four neurons are placed in the output layer since it is a four-class (pertussis, bronchitis, normal and COVID-19) classification model. The cepstral analysis is performed on the Mel spectrums of cough sound dataset to extract Mel Frequency Cepstral Coefficients (MFCC) (S. Gupta & Suhel, 2015). Principal Components Analysis (PCA) is employed on these MFCC to excerpt low dimensional features. The CML-MC uses an SVM classifier with a k-fold validation technique on combined (average MFCC and top few principal components) input data (Imran et al., 2020).

AI4COVID-19 mobile application computation engine predictions, i.e., "COVID-19 liable" when the person is not infected from COVID-19 or vice-versa, are tremendously low during the validation phase. The cough detection phase showed an accuracy of 95.60%, while the Cough diagnosis phase showed an accuracy of 92.85% (DTL-BC), 92.64% (DTL-MC) and 88.76% (CML-MC), respectively. All the classification metrics can be further improved with the availability of more data (Imran et al., 2020).

10.9 CONCLUSION AND FUTURE SCOPE

The AI domain of science is undergoing a golden era for the past two decades. AI has drastically enhanced the existing mobile applications with many features like personalization, error reduction, future prediction, real-time assistance, data ingestion, mimics human cognition, better data handling and management. AI-powered mobile applications are performing outstandingly well with high accuracy metrics. In this chapter, we have discussed many such remarkable applications, from health care to agriculture. AI algorithms are deployed into mobile apps to accomplish the following functionalities; Predicting the onset of stroke and melanoma promotes early treatment. Skin infections are easily classified based on their class labels without any physical interaction with the patient. Plant leaf disease is categorized to correctly identify the disease-causing pathogen, thus reducing the chances of losing the crop. Efficient management of Hypoglycaemia condition is assured by predicting the state of the patient during the critical event. Precisely locating the implanted CIEDs present in the patient body through chest radiograph analysis is very useful during an emergency. COVID-19 infected person identification using cough sound is a pragmatic approach that can be combined with temperature scanning.

AI models, especially DL models, are always data-hungry and their performance is enhanced on the increase with patient records. In the future, the mobile application accuracy of prediction is assured by growing records accumulated during initial trials. Most of the pre-trained models are meant for a high computation environment. Implementing a light version of such architectures along with a transfer learning approach would improve the diagnostics capabilities of the mobile application. Deployment and support for high computation AI algorithms using Cloud technologies can effectively address the implementation hurdles of smartphone apps.

REFERENCES

Achanta, R., Shaji, A., Smith, K., Lucchi, A., Fua, P., & Süsstrunk, S. (2012). SLIC Superpixels Compared to State-of-the-Art Superpixel Methods. *IEEE Transactions on Pattern Analysis and Machine Intelligence*, *34*(11), 2274–2282. https://doi.org/10.1109/TPAMI.2012.120

Ahad, M. A., Paiva, S., Tripathi, G., & Feroz, N. (2020). Enabling technologies and sustainable smart cities. *Sustainable Cities and Society*, *61*, 102301. https://doi.org/10.1016/j.scs.2020.102301

Arevalo-Rodriguez, I., Smailagic, N., Roqué, I., Figuls, M., Ciapponi, A., Sanchez-Perez, E., Giannakou, A., Pedraza, O. L., Bonfill Cosp, X., & Cullum, S. (2015). Mini-Mental State Examination (MMSE) for the detection of Alzheimer's disease and other dementias in people with mild cognitive impairment (MCI). *The Cochrane Database of Systematic Reviews*, *2015*(3), CD010783–CD010783. https://doi.org/10.1002/14651858.CD010783.pub2

Bergier, H., Duron, L., Sordet, C., Kawka, L., Schlencker, A., Chasset, F., & Arnaud, L. (2021). Digital health, big data and smart technologies for the care of patients with systemic autoimmune diseases: Where do we stand? *Autoimmunity Reviews*, *20*(8), 102864. https://doi.org/10.1016/j.autrev.2021.102864

Blair Ian. (2021). *Mobile App Download and usage Statistics (2021)*. Buildfire. https://buildfire.com/app-statistics/

CameraX API. (2021). https://developer.android.com/training/camerax

Chollet, F. (2015). Keras. In *GitHub repository*. GitHub. https://github.com/fchollet/keras

Cohen, J. (2020). *Not wearing masks to protect against coronavirus is a 'big mistake,' top Chinese scientist says*. SciencemageOrg.

Duan, L., & Xu, L. Da. (2012). Business Intelligence for Enterprise Systems: A Survey. *IEEE Transactions on Industrial Informatics*, *8*(3), 679–687. https://doi.org/10.1109/TII.2012.2188804

Ech-Cherif, A., Misbhauddin, M., & Ech-Cherif, M. (2019). Deep Neural Network Based Mobile Dermoscopy Application for Triaging Skin Cancer Detection. *2019 2nd International Conference on Computer Applications Information Security (ICCAIS)*, 1–6. https://doi.org/10.1109/CAIS.2019.8769517

Goceri, E. (2021). Diagnosis of skin diseases in the era of deep learning and mobile technology. *Computers in Biology and Medicine*, *134*, 104458. https://doi.org/10.1016/j.compbiomed.2021.104458

Goodfellow, I., Bengio, Y., & Courville, A. (2016). *Deep Learning*. MIT Press. http://www.deeplearningbook.org

Gupta, A. (2019). StrokeSave: A Novel, High-Performance Mobile Application for Stroke Diagnosis using Deep Learning and Computer Vision. *ArXiv E-Prints*. https://arxiv.org/abs/1907.05358

Gupta, S., & Suhel, M. (2015). Speech Recognition using MFCC \& VQ. *International Journal of Scientific Engineering And Technology Research*, *4*(01), 58–61.

He, K., Zhang, X., Ren, S., & Sun, J. (2016). Deep residual learning for image recognition. *Proceedings of the IEEE Conference on Computer Vision and Pattern Recognition*, 770–778.

Horsham, C., Loescher, L. J., Whiteman, D. C., Soyer, H. P., & Janda, M. (2016). Consumer acceptance of patient-performed mobile teledermoscopy for the early detection of melanoma. *British Journal of Dermatology*, *175*(6), 1301–1310. https://doi.org/10.1111/bjd.14630

Howard, A. G., Zhu, M., Chen, B., Kalenichenko, D., Wang, W., Weyand, T., Andreetto, M., & Adam, H. (2017). Mobilenets: Efficient convolutional neural networks for mobile vision applications. *ArXiv Preprint ArXiv:1704.04861*.

Howard, J. P., Fisher, L., Shun-Shin, M. J., Keene, D., Arnold, A. D., Ahmad, Y., Cook, C. M., Moon, J. C., Manisty, C. H., Whinnett, Z. I., Cole, G. D., Rueckert, D., & Francis, D. P. (2019). Cardiac Rhythm Device Identification Using Neural Networks. *JACC: Clinical Electrophysiology*, *5*(5), 576–586. https://doi.org/10.1016/j.jacep.2019.02.003

Imran, A., Posokhova, I., Qureshi, H. N., Masood, U., Riaz, M. S., Ali, K., John, C. N., Hussain, M. D. I., & Nabeel, M. (2020). AI4COVID-19: AI enabled preliminary diagnosis for COVID-19 from cough samples via an app. *Informatics in Medicine Unlocked*, *20*, 100378. https://doi.org/10.1016/j.imu.2020.100378

Jacob, S., Shahzad, M. A., Maheshwari, R., Panaich, S. S., & Aravindhakshan, R. (2011). Cardiac Rhythm Device Identification Algorithm using X-Rays: CaRDIA-X. *Heart Rhythm*, *8*(6), 915–922. https://doi.org/10.1016/j.hrthm.2011.01.012

Lu, Y. (2019). Artificial intelligence: a survey on evolution, models, applications and future trends. *Journal of Management Analytics*, *6*(1), 1–29. https://doi.org/10.1080/23270012.2019.1570365

Martín, A., Ashish, A., Pau, B., Eugene, B., Zhifeng, C., Craig, C., Greg, S.C., Andy, D., Jeffrey, D., Matthieu, D., Sanjay, G., Ian, G., Andrew, H., Geoffrey, I., Michael, I., Jia, Y., Rafal, J., Lukasz, K., Manjunath, K., … Xiaoqiang, Z. (2015). *{TensorFlow}: Large-Scale Machine Learning on Heterogeneous Systems*. https://www.tensorflow.org/

Miranda, I. D. S., Diacon, A. H., & Niesler, T. R. (2019). A Comparative Study of Features for Acoustic Cough Detection Using Deep Architectures*. *2019 41st Annual International Conference of the IEEE Engineering in Medicine and Biology Society (EMBC)*, 2601–2605. https://doi.org/10.1109/EMBC.2019.8856412

Mohamed Yousuff, A. R., & Rajasekhara Babu, M. (2020). Improving the Accuracy of Prediction of Plant Diseases Using Dimensionality Reduction-Based Ensemble Models. *Emerging Research in Data Engineering Systems and Computer Communications*, 121–129. https://doi.org/10.1007/978-981-15-0135-7_11

NHLBI. (2019). Stroke. In *National Institute of Health*.

Oerke, E.-C. (2006). Crop losses to pests. *The Journal of Agricultural Science*, *144*(1), 31–43. https://doi.org/10.1017/S0021859605005708

Organization, W. H., & others. (2020). *Report of the WHO-China joint mission on coronavirus disease 2019 (COVID-19)*.

Pagiatakis, C., Rivest-Hénault, D., Roy, D., Thibault, F., & Jiang, D. (2020). Intelligent interaction interface for medical emergencies: Application to mobile hypoglycemia management. *Smart Health*, *15*, 100091. https://doi.org/10.1016/j.smhl.2019.100091

Pan, S. J., & Yang, Q. (2010). A Survey on Transfer Learning. *IEEE Transactions on Knowledge and Data Engineering*, *22*(10), 1345–1359. https://doi.org/10.1109/TKDE.2009.191

Picon, A., Alvarez-Gila, A., Seitz, M., Ortiz-Barredo, A., Echazarra, J., & Johannes, A. (2019). Deep convolutional neural networks for mobile capture device-based crop disease classification in the wild. *Computers and Electronics in Agriculture*, *161*, 280–290. https://doi.org/10.1016/j.compag.2018.04.002

Piczak, K. J. (2015). ESC: Dataset for Environmental Sound Classification. *Proceedings of the 23rd ACM International Conference on Multimedia*, 1015–1018. https://doi.org/10.1145/2733373.2806390

Ratzki-Leewing, A., Harris, S. B., Mequanint, S., Reichert, S. M., Belle Brown, J., Black, J. E., & Ryan, B. L. (2018). Real-world crude incidence of hypoglycemia in adults with diabetes: Results of the InHypo-DM Study, Canada. *BMJ Open Diabetes Research and Care*, *6*(1). https://doi.org/10.1136/bmjdrc-2017-000503

Rogers, D., Campbel, R., Catha, G., Patterson, K., & Puccio, D. (2016). *Arrhythmia*. American Heart Association. https://www.heart.org/en/health-topics/arrhythmia/about-arrhythmia

Sandler, M., Howard, A., Zhu, M., Zhmoginov, A., & Chen, L.-C. (2018). MobileNetV2: Inverted Residuals and Linear Bottlenecks. *2018 IEEE/CVF Conference on Computer Vision and Pattern Recognition*, 4510–4520. https://doi.org/10.1109/CVPR.2018.00474

Simionescu, C., Insuratelu, M., & Herscovici, R. (2020). Prehospital Cerebrovascular Accident Detection using Artificial Intelligence Powered Mobile Devices. *Procedia Computer Science*, *176*, 2773–2782. https://doi.org/10.1016/j.procs.2020.09.279

Singh, V., & Misra, A. K. (2017). Detection of plant leaf diseases using image segmentation and soft computing techniques. *Information Processing in Agriculture*, *4*(1), 41–49. https://doi.org/10.1016/j.inpa.2016.10.005

SkinCareFoundation. (2021). *Skin Cancer Facts & Statistics*. https://www.skincancer.org/skin-cancer-information/skin-cancer-facts/

Staff, B. co. (2014). Medical gallery of Blausen Medical 2014. *WikiJournal of Medicine*. https://doi.org/10.15347/WJM/2014.010

Staff, B. R. (2019). *Wearable Medical Devices: Technologies and Global Markets. Business Communications Company*. https://www.bccresearch.com/market-research/healthcare/wearable-medical-devices.html

Torai, S., Chiyoda, S., & Ohara, K. (2020). Application of AI Technology to Smart Agriculture: Detection of Plant Diseases. *2020 59th Annual Conference of the Society of Instrument and Control Engineers of Japan (SICE)*, 1514–1519. https://doi.org/10.23919/SICE48898.2020.9240353

Tripathi, G., Ahad, M. A., & Paiva, S. (2020a). SMS: A Secure Healthcare Model for Smart Cities. *Electronics*, *9*(7). https://doi.org/10.3390/electronics9071135

Tripathi, G., Ahad, M. A., & Paiva, S. (2020b). S2HS- A blockchain based approach for smart healthcare system. *Healthcare*, *8*(1), 100391. https://doi.org/10.1016/j.hjdsi.2019.100391

van Doremalen, N., Bushmaker, T., Morris, D. H., Holbrook, M. G., Gamble, A., Williamson, B. N., Tamin, A., Harcourt, J. L., Thornburg, N. J., Gerber, S. I., Lloyd-Smith, J. O., de Wit, E., & Munster, V. J. (2020). Aerosol and Surface Stability of SARS-CoV-2 as Compared with SARS-CoV-1. *New England Journal of Medicine*, *382*(16), 1564–1567. https://doi.org/10.1056/NEJMc2004973

Weinreich, M., Chudow, J. J., Weinreich, B., Krumerman, T., Nag, T., Rahgozar, K., Shulman, E., Fisher, J., & Ferrick, K. J. (2019). Development of an Artificially Intelligent Mobile Phone Application to Identify Cardiac Devices on Chest Radiography. *JACC: Clinical Electrophysiology*, *5*(9), 1094–1095. https://doi.org/10.1016/j.jacep.2019.05.013

WorldStrokeOrganization. (2019). *Stroke Annual Report*. WSO. https://www.world-stroke.org/assets/downloads/WSO_2019_Annual_Report_online.pdf

Yale, J.-F., Paty, B., & Senior, P. A. (2018). Hypoglycemia. *Canadian Journal of Diabetes*, *42*, S104–S108. https://doi.org/10.1016/j.jcjd.2017.10.010

11 Role of Emerging Technologies in Energy Transformation and Development of Clean and Green Energy Solutions

Sunil Kumar Khare
University of Petroleum and Energy Studies, Dehradun, India

CONTENTS

DOI: 10.1201/b23013-11

11.1 INTRODUCTION

Energy is the basic need for survival of humans. Life could survive on earth due to abundant availability of sunlight and fire, the two primary natural sources of energy. The demand of energy increased with advancement of civilization. During industrial revolution, coal became main energy source for running factories. In the middle of nineteenth century, hydrocarbon discovery and its usage in internal combustion and jet engines revolutionized the transport sector. People could travel much faster on road, in air and in sea, which led to businesses grow manifold. The logistics and supply chain of products at global scale led to flourishing multinational corporations supplying goods and services across the world. The mobilization of people around the globe became a common phenomenon due to technology development and flourishing services sector.

In later years, technologies developed for commercial use of nuclear, solar, wind, tidal and geothermal energy. Energy usage became enabler of human growth. In fact, the per capita energy consumption became a measure of economic growth and prosperity of society and nations. While western economies became major consumers of energy, the Asian and African economies lagged behind in per capita energy consumption. The end of twentieth century saw the rising demand of energy from developing and highly populous Asian countries like India and China. The global energy consumption has seen a steady rise in last few decades and we see no signs of its waning.

Due to continuous increase in global energy demand and with limited conventional sources of energy, we have to think of innovative ways to increase our energy production and decrease its consumption without making major impact on our lifestyle, which is largely energy dependent. According to United States of America (USA) energy information Administration data, there is an increase in USA energy consumption from 35 to 101 Quadrillion British thermal units (Btu), from year 1950 to 2019. The share of fossil fuel in year 2019 was around 80 Quadrillion while nuclear and renewables was 8 and 11 Quadrillion Btu, respectively. According to Energy international reports, the total energy consumption of India increased from 440 Mega ton oil equivalent (Mtoe) in year 2000 to 880 Mtoe in year 2020. Coal remains the primary energy source with 44% share while oil and natural gas contribute 31% to energy supply. The share of renewables in energy supply is meagre 3% while bio mass contribute 13%.

A major challenge in use of fossil fuels is the air pollution. Constant discharge of carbon dioxide (CO_2) gas in the atmosphere has led to warming of atmosphere (Dunne et al., 2013). Scientists predict 2 degree centigrade rise in global average atmospheric temperatures by end of twenty first century, which will lead to melting of some polar ice bodies and increase in the mean sea level on earth, eventually leading to transgression of seas and several cities submerging into sea water. Therefore, there is a thrust to reduce consumption of fossil fuel and move towards clean energy sources. In the post COVID world, a paradigm shift towards clean and green sources of energy is inevitable around world. Several countries are working to develop green hydrogen technologies. There is emphasis on development of solar parks to generate electricity. Electrical vehicles are in demand now. The transition from non-renewable fossil fuel to sustainable clean renewable energy is bound to take place.

Research and analysis is going on in Europe to transition to 100% clean and renewable energy by year 2050 (Child et al., 2019). The transition of energy will be from dominant fossil fuel to dominant renewables (Figure 11.1). Last few decades have seen growth of knowledge economy and digital technologies, which have transformed the way we think and live. The emerging digital technologies will expedite our energy transition by helping develop innovative smart solutions and products.

In this chapter, we discuss in detail the current energy outlook and its transition in near future. We discuss the share of fossil fuels like oil, natural gas and coal in current energy basket. We discuss the role of digital technologies towards optimization of exploration and production of oil, gas and coal. We also discuss the increasing role of unconventional hydrocarbons like coal bed methane (CBM), shale gas and natural gas hydrates (NGH) in total energy basket. Next, we discuss the current state of renewables and technology trends in their commercial development as sustainable and clean energy sources. We also discuss the role of emerging digital technologies as enablers of energy transition to clean and green energy.

The digital technologies like big data, analytics, and internet of things (IoT), artificial intelligence (AI) and machine learning (ML) (Figure 11.2) help us develop innovative solutions and products that help us move from fossil fuels to clean renewable

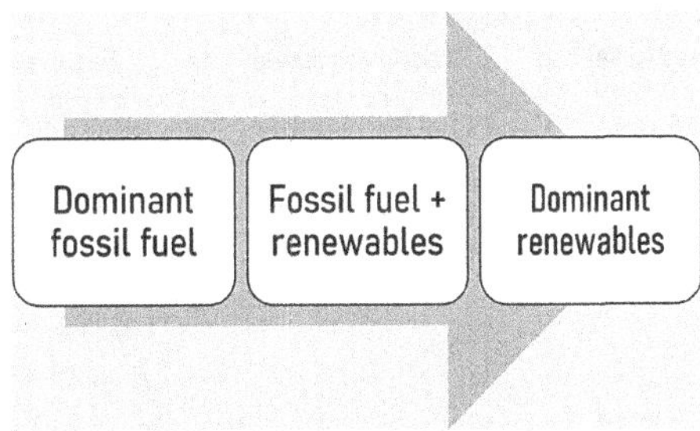

FIGURE 11.1 Future trends in energy production and consumption with transformation from present dominant fossil fuel to future dominant clean and sustainable energy sources. This transition will be over by year 2050.

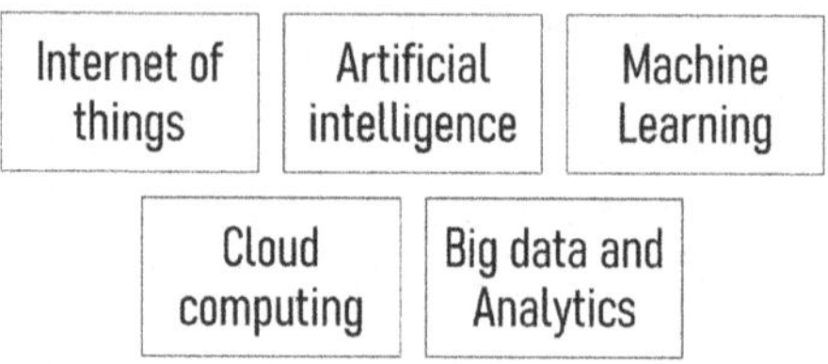

FIGURE 11.2 The emerging digital technologies, which can affect our energy production and usage pattern. Optimization of clean energy production with the help of Artificial intelligence, machine learning, Internet of things, cloud computing and data analytics technologies can fast track energy transition.

energy. The necessity to decrease carbon footprint due to threats of global warming has led to development of clean energy technologies, for which the digital technologies are big enablers. Next gen technologies help us increase clean energy production.

The emerging technologies help control energy losses and decrease energy consumption. Several innovative technologies are there to produce clean energy from coal seams, shale formations, gas hydrates, hydrogen, sea tides, solar and geothermal sources etc. In this chapter, we describe major sources of energy, its consumption pattern and try to understand technology led interventions, which enhance energy production.

11.2 ENERGY SOURCES

We use several energy sources to meet our energy demand. Majority of energy consumption is in industries, transport and households. While coal is the prime ingredient for electricity generation and for heating industrial furnaces, we need hydrocarbons in transport sector for running engines of motor vehicles, ships and aircrafts. Households also use hydrocarbons for cooking, heating, and generation of electricity.

To meet our energy demand, we use several types of energy sources, broadly classified into three categories shown in Figure 11.3.

In following sections, we will look at each energy source and discuss in detail the role of emerging technologies in their production and use.

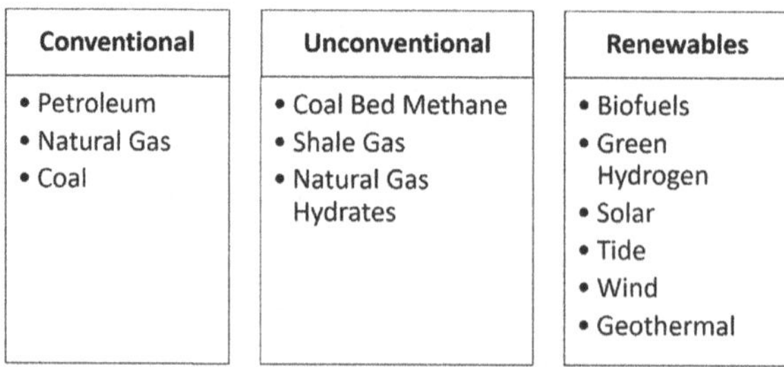

FIGURE 11.3 The energy sources on earth clubbed as conventional, unconventional and renewable energy sources. The conventional and unconventional are fossil fuels while renewables production takes place from biomass and energy of the nature.

11.3 CONVENTIONAL ENERGY SOURCES

Oil, Gas and coal are the three most important conventional energy sources. They are fossil fuels because their production from dead organic matter deposited in sedimentary basins. They are conventional, because of their production from conventional sandstone, limestone and dolomite subsurface reservoirs. They have been powering our homes and machines for more than a century. In below section, we discuss in brief about conventional energy sources and role of emerging technologies in their production.

11.3.1 Oil and Gas

Oil and gas is produced from sedimentary basins after intensive phase of exploration and development of oil and gas field. The exploration and production of oil and gas is capital-intensive business, which involves application of high-end technologies in inhospitable terrains. The success rate of striking oil and gas in a drilled well is low; the global average stands at 34% in year 2015–2019. Research and development was done to improve the well success rate. Petroleum reserves are located subsurface, several thousand feet below ground in both land and offshore basins. The logistics and mobilization of people to work in these inhospitable locations is a challenge. High pore pressure of petroleum reservoir is balanced by drilling fluids during construction of wellbore. Health and safety of people and environmental compliances are key areas of concern during recovery of hydrocarbons from subsurface. Production of high viscosity crude oil from subsurface reservoir is a technological challenge. The percentage recovery of hydrocarbon from subsurface reserves is still low; the global average stands at 30–35% initial oil in place (IOIP) in reservoir. The average gas recovery from a shale gas reservoir is 20–25% while shale oil recovery is less than 10%. Therefore, implementation of right techniques of Enhanced Oil Recovery (EOR) is an area of priority for optimum recovery of oil and gas from reservoirs. The role of emerging technologies assumes high importance due to consistent low crude price, which is on average less than 75 US dollar (USD) per barrel since year 2015. The new oil reservoirs are now hard to discover and produce due to depletion of easy access oil. They are located in deep, ultradeep waters and in inhospitable terrains. The broad areas of upstream operations where emerging technologies have big role to play is shown in Figure 11.4.

Below section describes the role of emerging technologies in petroleum exploration and production.

11.3.1.1 Emerging Technologies for Petroleum Exploration and Production

The exploration and production of petroleum is a data driven modelling and simulation approach (Koroteev and Tekic, 2021). The petroleum reserves are located several thousand feet below ground with no direct access to assess the extent and characteristics of reservoir. Seismic, gravity and resistivity data of subsurface rocks is generated to infer the presence of oil and gas. The data volume is large, whose storage, processing and handling has traditionally remained a challenge. Now with the advent of big data technology and presence of servers on cloud, the storage and

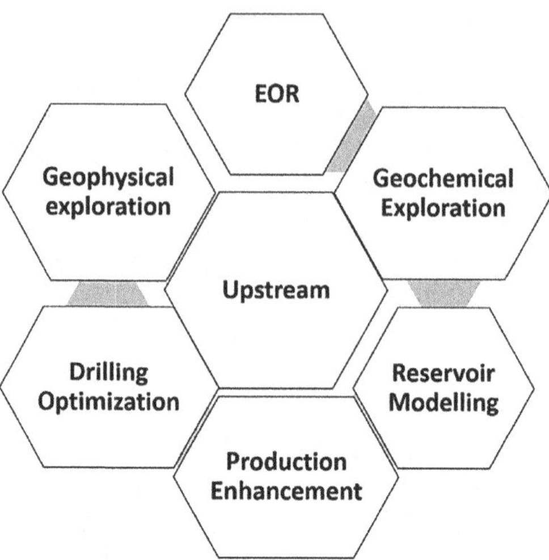

FIGURE 11.4 The areas in upstream exploration and production where emerging technologies can optimize operations.

quality control (QC) of data has become more efficient. Data analytics engines help remove spikes and noises by removing bad signals and thereby enhancing the data quality. Simultaneous handling of large data volumes requires a big data platform (Mohammadpoor and Torabi, 2020). Hadoop, MongoDB, Teradata, SAP HANA, Microsoft Azure, Amazon web services are some big data platforms widely used for storage and processing of oil and gas upstream data.

Another area of importance where emerging technologies have major role to play is drilling and its automation (Figure 11.5). Drilling of reservoir is a high risk, capital-intensive project where there are vast opportunities for improvement. Managed pressure drilling (MPD) and underbalanced drilling (UBD) apply choke manipulation techniques to increase or decrease the equivalent circulating density (ECD) of the drilling fluid in the well, thereby managing formation pressure without actually changing the drilling fluid density. Addition of nanoparticles in drilling fluids helps manage their properties to meet rheological and density issues. Logging while drilling (LWD) has enabled real time logging of the borehole during its construction phase. It saves valuable rig time and enables real time decision related to wellbore quality. Rotary steering has enabled better steering and control over deviated and horizontal wells. Casing while drilling is another emerging technology where well casing is run along with drill string so that the wellbore is cased while being drilled. Several new wellbore logging tools based on nuclear magnetic resonance (NMR), neutron density, resistivity and gamma ray technologies are run to better understand Petrophysical properties of the subsurface reservoir rocks.

Real time drilling data is generated from several sensors located on drill string and rig equipment. These sensors generate real time data of bit depth, hole depth, Weight on bit (WOB), revolutions per minute (RPM), stand pipe pressure (SPP), formation gas, mud volumes, mud density, mud temperature, well head pressure (WHP) and

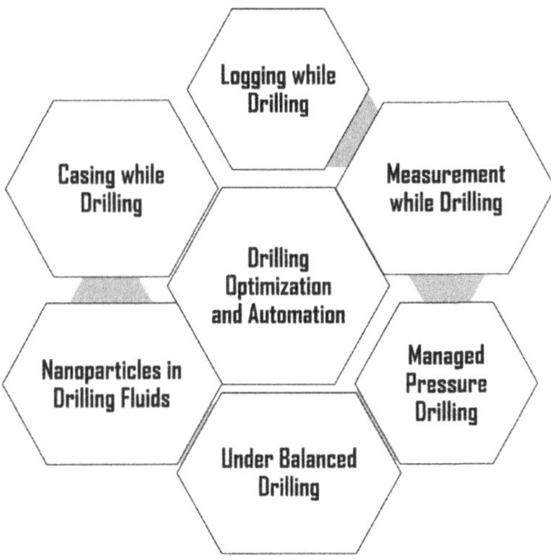

FIGURE 11.5 The emerging technologies in oil and gas drilling domain, which enable drilling optimization and automation.

several other critical variables. Mud logging units deployed at wellsite generate these drilling, mud and gas related data in real time. They display this data in dashboards and real time graphics, used by key personnel at wellsite for drilling decision support. Analytics engine run on such real time drilling data predicts and mitigates drilling hazards like stuck pipe, string wash out, mud motor failure and bit wear. Analytics engine also recommends predictive maintenance of equipment and critical machines on rig, which helps optimize supply chain and logistics related to spares and maintenance of rig equipment. Emerging drone technologies can perform rig inspection of areas, which are hard to access manually. Locations on rig like mast, crown and other high-rise areas like monkey board need regular inspection for their integrity. In offshore drill ships, semi subs, barge and jack up rigs, manual inspection of hull, legs and pontoons remains a challenge due their inaccessibility and their location towards the sea. The drone with high-resolution camera is handy and provides a clear view of these areas to the inspection engineers. In hazardous areas with high concentration of poisonous gases like hydrogen sulphide (H_2S) and CO_2, deploy robots for operation support. Video analytics is another emerging technology, deployed for operation surveillance on rigs. Deploy high-resolution cameras at strategic locations on the rig to monitor real time operations. The video stream thus generated helps get important insight on anomalies in operations. Infrared cameras assist in safety surveillance of production sites by detecting leaks of inflammable gases in the atmosphere and thereby raising an alarm to control the mishap. Drilling is a hazardous operation done in inaccessible terrains, therefore its automation and deployment of robots at rigs can lead to drilling efficiency optimization. Send the real time data generated at drill site to real time operation centre (RTOC) and regional operation centres (ROC), which are decision support centres located in the base and operated by senior engineers of the operating company. Bidirectional transmission of data and information from rig

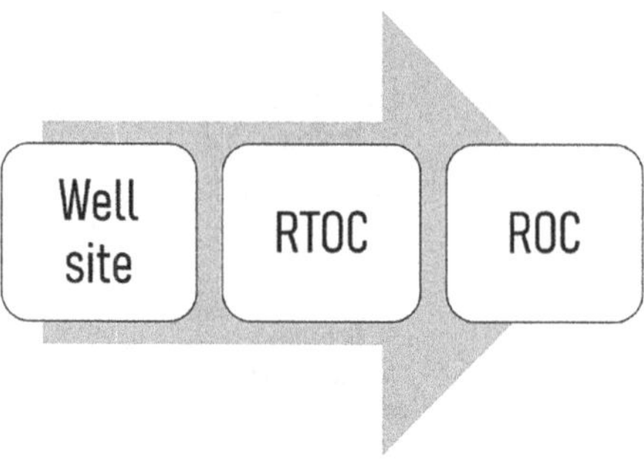

FIGURE 11.6 RTOC and ROC are monitoring and intervention stations in drilling operation. They operate 24 × 7 from base, manned by senior operation experts.

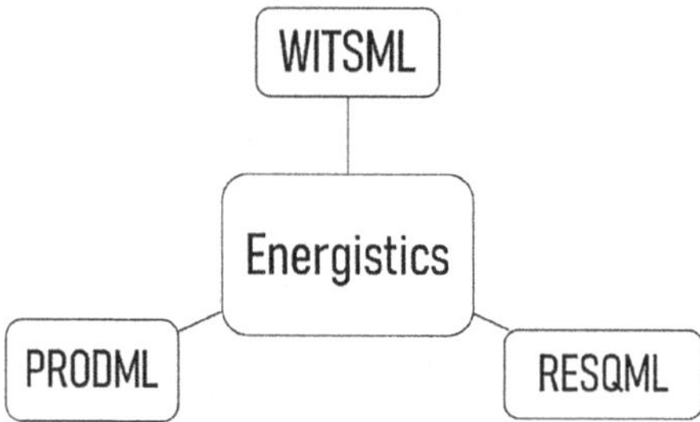

FIGURE 11.7 Drilling, reservoir and production data storage and transmission protocol developed by Energistics, which is a not for profit consortium of experts drawn from several oil and gas companies.

to RTOC and ROC and vice versa is the key feature of this system, which enables intervention of key experts in resolution and mitigation of drilling hazards and optimization of drilling operations (Figure 11.6). Implementation of machine learning and machine vision for automation of drilling operations can optimize the drilling operations (Khare, 2019, 2021)

Oil companies use wellsite information transfer standard markup language (WITSML), production markup language (PRODML) and reservoir markup language (RESQML) standard data transfer and storage protocols for transfer and storage of drilling, production and reservoir data, respectively, from well site to real time centres and to other stakeholders (Figure 11.7).

Energistics is a not for profit consortium of Oil and gas industry personnel who develop and update these data transmission and storage protocols (c.f. Energistics.org). Several oil and gas companies, oil field service companies and other technology companies actively participate in the development of this protocol.

The digital oilfields (DOF) are integrated network of oil producing wells of an oil field, with central command and control centres to meet the objective of desired production from the field. DOF involves implementation of digital technologies to optimize production with reduced risks and cost. These smart oil fields help improve the efficiency of oil fields. Intelligent oil fields use digital gadgets and tools, which are

IoT devices connected to Wi-Fi network of the company where real time information is sent to all stakeholders.

The petroleum reservoir simulation and modelling helps understand the aerial extent and characteristics of reservoir. Several oil field service companies like Schlumberger and Halliburton have their proprietary software application like Petrel and Landmark for real time simulation of reservoirs. These software use core, lithology, well test pressure and several other subsurface data sets to generate a real time simulation of the reservoir. The emerging digital technologies like AI and ML can help review and update reservoir models, which are more reliable.

11.3.2 Coal

Coal is fossil fuel produced from opencast and underground mines (Figure 11.8).

Opencast coal mining refers to a surface mining technique, which extracts coal by making pits from surface of earth to the coal seam. In this technique the overburden soil and rock mass is removed to reach the coal seam. The opencast mines are dug in coalfields where the coal seam is located near surface (Chaulya and Prasad 2016). It is relatively simple to execute technique than long wall technique in which tunnel is dug into the earth to reach coal / ore body. Long wall technique is useful where coal seams are located several thousand feet below surface of earth. Surface mining of coal requires use of detonators, excavators, trucks and dumpers to separate overburden and remove coal from pits. Long wall mining requires digging of tunnels and then extraction of coal from underground seams to surface of earth.

11.3.2.1 Emerging Technologies in Coal Sector

Emerging technologies are very beneficial in exploration and production of coal. Commercial software like SURPAC are available for modelling and simulation of reserves from core and other related data. The software has 3D graphics, which help

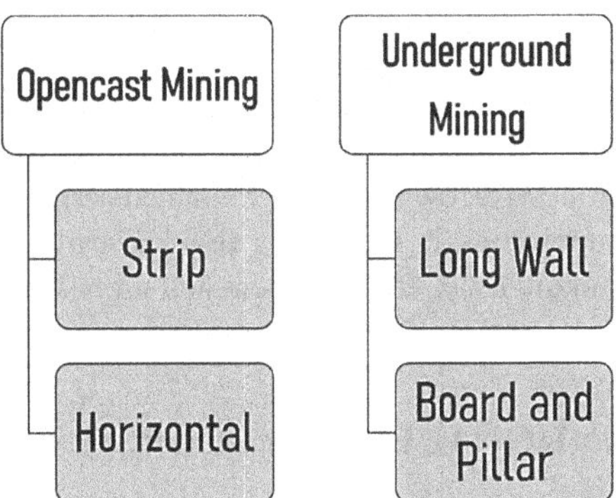

FIGURE 11.8 Various surface and subsurface mining methods used for extraction of coal. The coal seams may be located close to surface or deep inside earth. The opencast mining techniques are used for shallow depth seams while underground mining techniques are used for seams located deep inside ground.

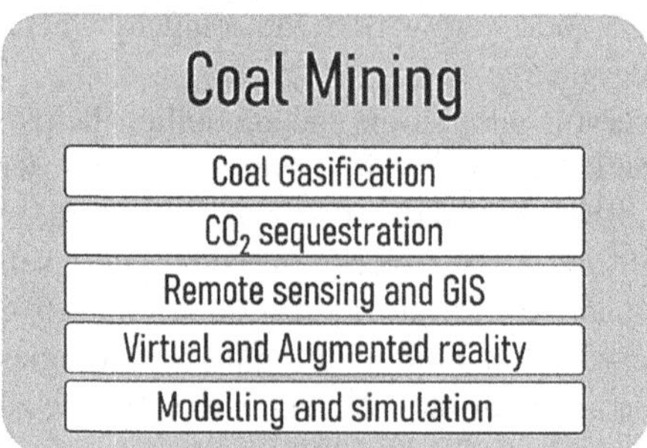

FIGURE 11.9 The engineering and digital technologies used to optimize coal mining and production of clean energy from coal.

understand 3-D aerial extent of coal seam. Automation of mining workflows can be achieved by input of data in workflows of this software. The global foremost use of coal is in electricity generation in thermal power plants. Burning of coal produces CO_2, which is responsible for global warming. Therefore, there is lot of emphasis on reduction of CO_2 in atmosphere. CO_2 sequestration technique have been developed to capture and store CO_2 generated as industrial effluent, and pump them inside coal seams to release trapped methane, which is a clean energy source.

Coal gasification is another technique to convert coal into combustible gas. CO_2 is removed from gas before it is burnt that helps restrict release of CO_2 in atmosphere It is an effective technique to reduce global warming. Digital technologies are deployed for safe mining operations, and to optimize coal production (Figure 11.9).

Technologies like virtual and augmented reality are used to train executives in a simulated environment. They also help better planning and simulation of mine in office premises. The operational hazards and their mitigation are explained to miners in an augmented reality enabled virtual environment.

Remote sensing and geographic information system (GIS) is useful to miners for geospatial information related to location and accessibility of coal body. AI and ML are widely used for predictive maintenance of machines, operational efficiency optimization and mitigation of mining hazards. Autonomous vehicles and self-driven trucks can navigate through narrow tunnels. Unmanned aerial vehicles are being used for surveillance and monitoring in hazardous areas. They are widely used for asset management. Close Circuit TV (CCTV) camera and drones are important safety and surveillance tools.

11.4 UNCONVENTIONAL ENERGY SOURCES

The hydrocarbons produced from unconventional reservoirs are unconventional fossil fuels. Most such reservoirs produce light hydrocarbons, and are promising energy source. Production of unconventional hydrocarbons started few decades ago. The two important unconventional hydrocarbon sources, which produce commercial quantity of energy, are coal bed methane (CBM) and shale gas.

11.4.1 Coal Bed Methane

The most common source of unconventional hydrocarbon is CBM, produced from coal seams by drilling a borehole through them, which enables release of trapped light hydrocarbons from the seams into the borehole (Haldar, 2018). CBM has less carbon footprint and thus environment friendly. Research is on to make them a viable commercial substitute of conventional hydrocarbons and an alternate energy source. Large quantities of methane gas produced from organic contents during coalification inside earth's crust. This gas remained trapped in coal seams and produced by drilling gas wells through coal bearing formations. The methane gas escapes into atmosphere when mining the coal seams, if the prior recovery of methane not done. A special technique called carbon dioxide (CO_2) sequestration recovers methane before mining of coal. In 2019, global CBM market was of USD 16 Billion, which grows at the rate of around 6% annually. An analysis done by Grand view research reveals that major consumption of CBM is in power generation (41 %), commercial (12.7 %), industries and residential sectors.

Russia has largest CBM reserves in world, estimated to be 83.7 trillion mt^3. USA CBM production in year 2017 was 980 billion cubic feet; with total CBM, reserves estimated around 11.88 trillion cubic feet.

India has about 2600 billion cubic meters CBM reserves spread in 12 states. As per reports of Govt. of India Directorate General Hydrocarbon (DGH), Oil and Natural Gas Corporation Limited (ONGC), Essar Oil and Reliance Industry Limited (RIL) are major companies producing CBM from coalfields in the Damodar Koel valley and Son valley in states of West Bengal and Jharkhand.

11.4.1.1 CO_2 Sequestration

The CO_2 sequestration optimizes methane recovery from coal seams. Boreholes are dug into the seams to recover trapped methane. CO_2 is pumped into the wells that leads to release of methane from coal seam and its production.

11.4.2 Shale Gas

Another promising source of unconventional hydrocarbon is shale gas. USA is the largest producer of shale gas in the world, with total production around 7.4 trillion cubic feet (TCF) in year 2019. However, China has the largest total recoverable reserves in the world. While majority of hydrocarbon produced from Shale formation is gas, some oil is also produced. Methane is the primary hydrocarbon produced from shale formations while ethane, propane and butane are produced in minor amounts. Some shale formations also produce CO_2 and H_2S in minor amounts. Marcellus, Eagle ford and Barnett shale are some of the highest gas producing shale formations in USA. While Marcellus shale is located in Pennsylvania, the Barnett and Eagle ford shales are located in Texas State. In India, Cambay, Gondwana, Krishna-Godavari, Cauvery and Assam-Arakan basins are prospective basins with reserves of shale gas. According to Energy Information and Administration (EIA), an energy agency of USA, India has around 584 TCF of shale gas and 87 billion barrels of shale oil in four sedimentary basins.

11.4.2.1 Emerging Technologies - Shale Hydraulic Fracturing

Shale formations are highly porous but impermeable. Their pores trap large volumes of hydrocarbons which cannot move due to low permeability, therefore are not producible (Neil et al., 2020). Hydraulic fracturing is an artificial method of permeability enhancement of shale formations by fracturing them with a fluid pumped under very-high pressure. The fractures propagate deep into the shale formation, and a proppant filled into the fractures to avoid their closure when the pressure diminishes. The proppant can be silica grains or nanoparticles with strength to hold high pressure and maintain optimum permeability of shale. The high-pressure fluid could be water or gases like Nitrogen. Hydraulic fracturing increases shale permeability manifold, which enables recovery of methane trapped in the pore spaces of shale (Fonseca, 2014).

Simulation of hydraulic fracturing is done with commercial software. Hydraulic fracturing produces large volumes of wastewater, whose safe disposal is an environmental concern. The treatment and disposal of wastewater requires newer technologies (Vengosh et al., 2013). Waterless hydraulic fracturing is a viable option. Cryogenic nitrogen is used as a substitute of water for hydraulic fracturing.

For commercial recovery of shale gas, long multilateral and horizontal wells are drilled through shale formations. The drilling of such wells is a technology challenge due to poor drill cutting lifting and maintenance of hydrostatic head. Geosteering assembly is widely used to steer the bit through horizontal shale sections. The shales are often sticky, friable and splintery. The use of appropriate drilling fluid system and maintenance of appropriate drilling parameters is key to success of geosteering horizontal wells. The sticky clay and shale in offshore environment poses challenge of stuck pipe due to differential sticking and pack offs. The use of appropriate synthetic mud systems and maintenance of lower drilling rates to avoid pack offs due to excessive cutting is mandatory for hassle free drilling of such sections. The drilled section is fractured with help of water or N_2 gas pumped at high pressure in the well. Once formation fractured, its permeability becomes manifold which provides channel for movement of low-density gaseous hydrocarbons to move towards wellbore. Proppant that are fine sand particles or Nano particles are pumped into the fractures to maintain the permeability for long time. The well testing and completion is done and gas is produced in a controlled manner.

11.4.3 Natural Gas Hydrates

Natural gas hydrates are an ice-like solid composed of water and gas, most commonly methane. They only form at high pressure and low temperatures, in places where both water and gas are plentiful. Small increase in temperature and decrease in pressure causes separation of water and gas. Different types of gas hydrates found in oceanic and artic environments are given in Figure 11.10.

Gas hydrates consist of molecules of natural gas (most commonly methane) enclosed within a solid lattice of water molecules (ice). When brought to the earth's surface, one cubic foot of gas hydrate can release 164 cubic feet of natural gas. Gas hydrate deposits are found wherever methane occurs in the presence of water under elevated pressures and at relatively low temperatures, such as beneath permafrost or in shallow sediments along deep-water continental margins. Once thought to be rare,

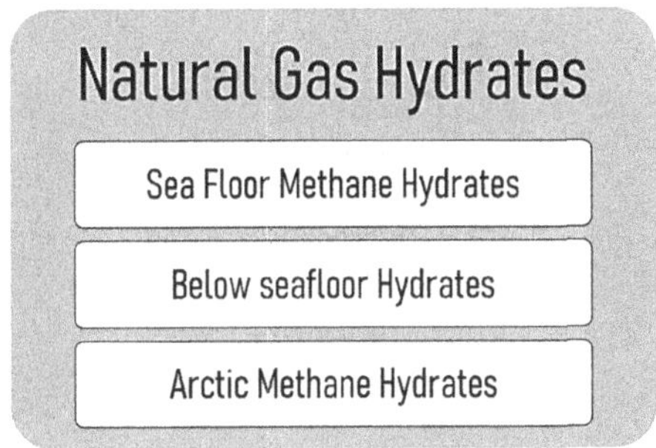

FIGURE 11.10 Different types of gas hydrates recovered from ocean and arctic permafrost environments.

gas hydrates occur in vast volumes and with global resource estimates from 250,000 to 700,000 trillion cubic feet (tcf) of natural gas compared to 2,829 tcf of technically recoverable resources of dry natural gas in the U.S.. India has huge reserves of natural gas hydrates spread along its shoreline spread from east to west (Sain and Gupta, 2012; Bhawangirkar et al., 2021). Advanced research work on two below given areas continues for commercial production of energy from natural gas hydrates.

11.4.3.1 Modelling and Analysis

Work includes simulation of field-scale production of gas from hydrates. It includes simulations for predicting geo-mechanical stability of hydrate-bearing sediment and hydrate-bearing reservoirs.

11.4.3.2 Fundamental Property Characterization of Hydrate Bearing Sediments

Work continues for experimental characterization to estimate geo-mechanical and relative permeability of hydrate reservoirs. They are numerical simulation inputs to simulate in situ conditions. Lab technique is being developed to form hydrates in laboratory-fabricated sediments.

11.5 NON-CONVENTIONAL RENEWABLE ENERGY SOURCES

Non-conventional energy sources are called renewable energy sources, as their reserves are infinite. They are produced from naturally occurring never-ending natural resources. The renewable energy sources are divided into categories shown in Figure 11.11.

In this section, we discuss the role of emerging technologies in each of these areas of non-conventional energy.

11.5.1 Biofuels

Biofuels are fuel derived from biomass. They are fuel of future and cater to need of transport sector (Nicolae et al. 2018). Technologies are developed for bio refining, biomaterials, environmental waste utilization, and biotechnology. Advanced research

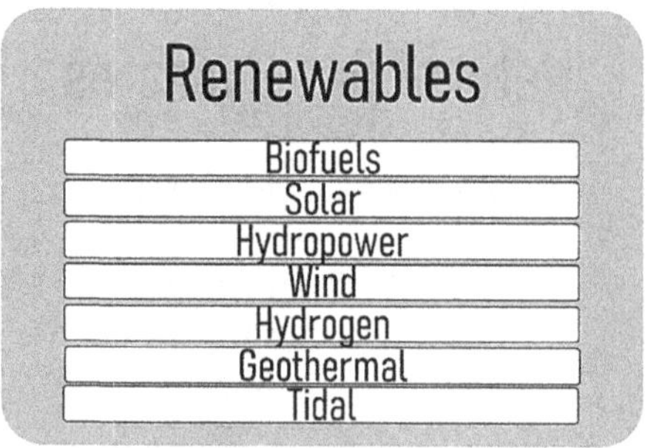

FIGURE 11.11 The renewable energy sources that can provide sustainable and clean energy. They are all clean and green energy sources, which do not pose to our environment.

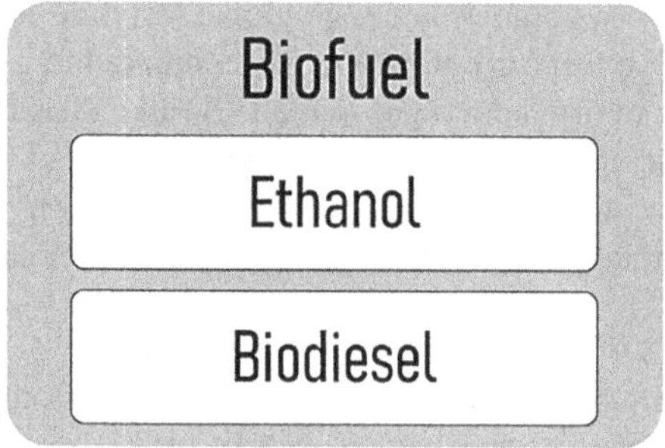

FIGURE 11.12 The classification of Biofuel into ethanol and biodiesel. While ethanol is made from plant starch, the biodiesel is derived from vegetable oil and animal fats.

is on to develop biofuels from non-food cellulosic and algae resources (Dhyani and Bhaskar, 2018; Chia et al., 2018). The common biofuels now available are biodiesel and ethanol (Figure 11.12), which are product of first generation biofuel technology.

Ethanol is made from plant starch and sugar materials, also called biomass. It acts as a blending agent for gasoline to increase octane and decrease harmful emissions. An average 10% ethanol is added to gasoline for blending. Fermentation of biomass results in ethanol formation.

Biodiesel is derived from vegetable oils and animal fats. It is used as substitute of diesel. It can be blended with petroleum-produced diesel to any proportions. Several companies worldwide are working towards production of bio diesel from waste fat, grease and oils (Ma et al., 2018). The biofuel conversion process involves either high temperature or low temperature deconstruction process followed by upgrading.

11.5.2 Hydrogen Fuel

Hydrogen is the most abundant element in solar system and the prime ingredient for fusion reaction in sun for solar energy generation. There is a global consensus on de-carbonization and energy transition, due to increase of greenhouse gases in atmosphere. Several governments are proposing green hydrogen initiatives to produce clean energy from hydrogen because it is environment friendly renewable energy source. Use of Hydrogen as industrial fuel to generate very high temperature heat is expected to commence commercially in next decade. The use of hydrogen as transport fuel is expected in next few years. With the initiatives of some biggest hydrogen project developers, the increase in green hydrogen production is expected to globally grow 50 fold in next six years.

The produced hydrogen is classified into green and grey hydrogen. Green hydrogen is different from grey hydrogen as it is produced by electrolysis to split water. Grey hydrogen is produced from methane resulting in release of greenhouse gases into the atmosphere.

Several initiatives are taken to reduce green hydrogen cost to less than 2 USD/kg. This will promote use of green hydrogen in most carbon-intensive industries with high emissions, e.g., Iron and steel, shipping, chemicals and power (Rand and Dell, 2008). Scaling up green hydrogen will be essential to help global economies to achieve net zero emissions by 2050 and limit global temperature rises to 1.5°C.

Abdalla et al. (2018) have done a review of challenges on application hydrogen as a fuel. Intervention of emerging technologies in green hydrogen sector has led to fall in their production costs 40% since 2015 that is expected to fall by a further 40% through 2025. Digital technology like AI and IoT can accelerate the transition to green hydrogen. Digital twins can help design models and perform their feasibility study. The emerging technologies for green hydrogen production are methane pyrolysis, solar hydrogen production, biological hydrogen production and biomass gasification (Figure 11.13).

We discuss in brief about these emerging hydrogen technologies.

11.5.2.1 Methane Pyrolysis

Plasma and thermal breakdown of methane produces hydrogen and carbon. The carbon thus produced is solid and stored. Hydrogen produced is a clean source of energy. Several companies are working on development of technologies for commercial scale methane pyrolysis. Researchers are working to optimize processes for hydrogen and carbon processes and removal of solid carbon from molten media.

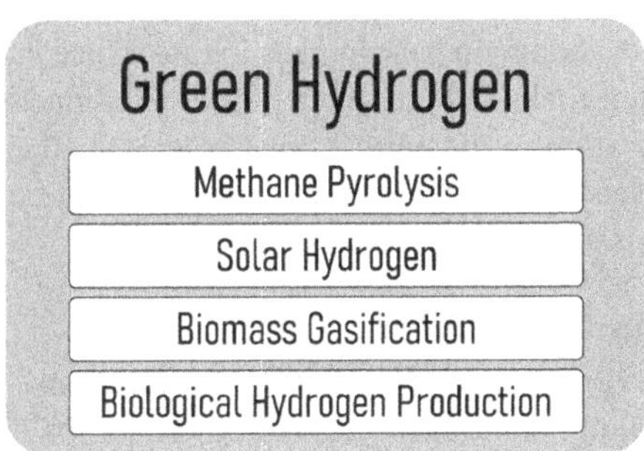

FIGURE 11.13 The four major emerging technologies in green hydrogen sector for commercial production of hydrogen as fuel.

11.5.2.2 Solar Hydrogen Production

This hydrogen production process uses solar energy to split water into hydrogen and oxygen. There are two ways to produce hydrogen from water, one photocatalytic and another photo-electrochemical water splitting (Alberto et al., 2021). Research is on to develop this technology for commercial production of hydrogen with least environment impact. The Photo-electrochemical water splitting (PEC) is used to generate hydrogen from water. It uses sunlight and specialized semiconductors called photo-electrochemical materials. These materials use light to dissociate water molecules.

11.5.2.3 Biological Hydrogen Production

Green algae uses sunlight and photolytic process to produce hydrogen. Several microbes produce hydrogen from their natural metabolic processes. Research is on to fully develop this technology for hydrogen production.

11.5.2.4 Biomass Gasification

Heat, steam and oxygen convert biomass to hydrogen and other products, without any combustion. Organic materials are converted to hydrogen at temperatures higher than 700°C. Biomass sources required for this conversion are agriculture and forestry crops, their residues, sewage, industrial residues, animal residues and municipal solid waste. This process of hydrogen production in recent year, development of electrolyser technology and its availability at economical price resulted in increase of green hydrogen production. The advent of green hydrogen production methods from methane pyrolysis and solar hydrogen production will open vista for widespread use of green hydrogen as renewable, clean, efficient and alternate energy source.

11.5.3 Geothermal Energy

Geothermal energy is derived from internal heat of earth. This energy is present in the fluid present in the pore spaces of rocks due to some anomalous heat source present in earth's crust. Geothermal energy sources have been broadly divided into low enthalpy and high enthalpy reservoirs (Kazmarczyk et al. 2020). Several workers have done detail investigation of geothermal provinces in India (Craig et al., 2013; Dimri 2013; Yadav and Sircar, 2021) and other parts of world (Gunnlaugsson, 2004, Carella, 1985). The hot water and steam gush out to surface due to their subsurface heating. One of the most common heat sources is subsurface magma chamber, which is a reservoir of molten rocks at high temperatures. Radioactive isotope decay and resultant energy generation in the earth's crust is also another important source of geothermal energy. Mantle convection results in transfer of heat from lower mantle to lithosphere, which is also a source of geothermal energy.

11.5.3.1 Technology Interventions

Geothermal wells are drilled to produce hot water and steam. Several drilling techniques like spark drilling, sonic drilling, hydrothermal spallation drilling, plasma drilling and MMW drilling are used to drill geothermal wells. Geothermal energy is used to generate electrical power. Remote sensing and geochemical techniques are used to explore geothermal energy sources. Seismic methods are also widely used for geothermal exploration. Engineered geothermal systems involve injection of water into dry impermeable rock in a region with high geothermal gradient. Two wells are drill around 100 m apart. Water is injected at high pressure form one well and hot water extracted from another nearby well.

Depleted oil and gas wells with high geothermal gradients and high bottom-hole temperatures can be ideal sites for geothermal energy generation. Water is injected through a water injection wells and the high temperature water is produced from nearby wells.

Geothermal steam at temperatures greater than 150°C drives a turbine to generate electricity. However, if steam temperature is less than 150°C, it is further heated to form dry steam, which is injected into a turbine to produce electricity.

Anderson and Rezaie (2019) have done an in depth analysis of geothermal technology, trends and its future role as sustainable energy source. Use of IoT and ML algorithms can automate the power generation equipment. Digital gadgets and apps can enable remote monitoring of geothermal plants. Data analytics can enable predictive maintenance of wells and power generation turbines and related equipment.

11.5.4 Tidal Energy

The tides are periodic motion of seawater induced by gravitational pull of moon and sun. The electricity is generated from tidal movement of seawater, which has periodic high and low movement called high and low tide, respectively. The high tide is associated with ascent of seawater with high energy, which causes turbines to move and generate electricity (Figure 11.14). The ascending water is stored in a manmade basin and released to rotate turbines during low tides when sea level is low. In this way, there is a cycle of power generation associated with high and low tide Tidal energy is renewable clean energy source, which if optimally developed can help meet our most of energy demand (Chowdhury et al. 2021).

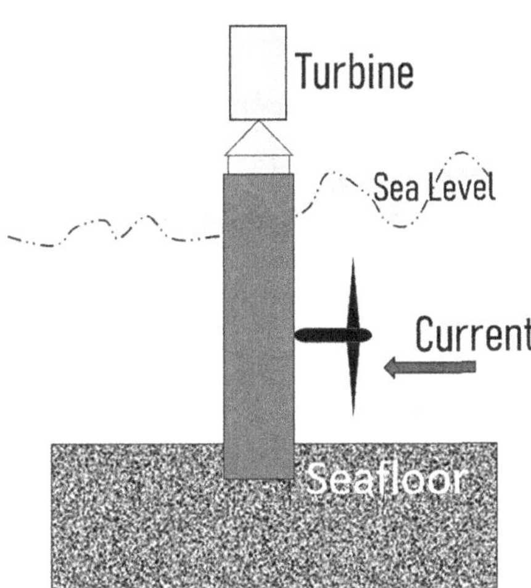

FIGURE 11.14 A turbine used to generate power from tidal waves.

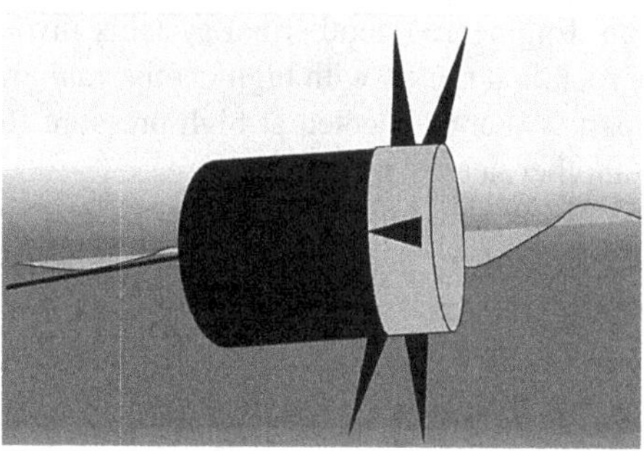

FIGURE 11.15 A second-generation tidal turbine.

11.5.4.1 Technology Interventions

Next generation tidal turbines have evolved towards full-scale commercial demonstration and show promise of substantively reducing the costs of tidal energy (Figure 11.15). The second-generation turbines have direct drive generators with no gearbox, passively yawing with the tidal current, therefore eliminating the need for active yaw or pitch control and use of a flexible mooring for station keeping (eliminating the need for a mono-pile). CoRMaT an example of such a system is the technology using two closely spaced, fixed pitch, contra-rotating rotors. These capture energy from the tidal flow to directly drive a contra-rotating generator. The development and performance of this prototype system has been reported extensively.

Monitoring the tidal energy assets for their integrity, performance and maintenance is a challenge in offshore areas (Uihlein and Magagna, 2016). Use of emerging digital technologies like video analytics with visual and infrared waves can capture real time data and transfer it to cloud server using client-server Wi-Fi technology (Johnstone et al. 2012).

11.5.5 Wind Energy

Energy of blowing wind has been used from time immemorial to generate power. Wind farms were developed in several countries to generate electricity. Wind farms have several collocated wind turbines used to produce electricity. In India, wind parks were commissioned at several locations (Sharma and Sinha, 2019) in Maharashtra and Rajasthan, with highest 25.5 MW power generation wind park located in Kanyakumari, Tamil Nadu.

11.5.5.1 Technology Interventions

Emerging construction, design and material technologies are deployed for efficiency gain in wind parks.

It includes deployment of taller towers and hubs (Figure 11.16). Hybrid towers made of steel and concrete are deployed to increase tower heights. Bigger blades are installed to harness more wind energy and to increase power generation. The need for

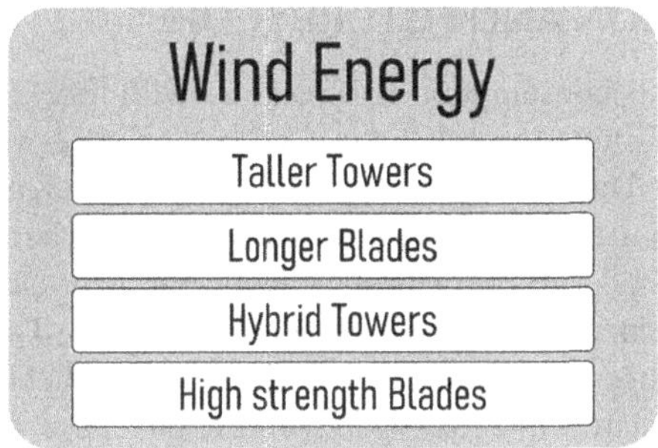

FIGURE 11.16 Possible optimization in wind turbine design that led to enhanced power generation in recent years.

longer blades has led to research and development of high-strength hybrid blade material. Reducing the mass of drivetrain components by innovative designs also helps in improving operational efficiency. AI and ML technologies can help monitor the condition of wind turbines (Stetco et al., 2019). Use of modular precast foundations is another emerging trend to add strength to tower and to fasten the commissioning of wind turbines.

11.5.6 Solar Energy

Sun is the primary energy source on earth. The existence of life on earth is due to constant supply of solar energy. Due to increase in demand of clean energy and associated environmental hazards of fossil fuel, solar is a reliable, cost effective and everlasting renewable energy source (Kannan and Vakeesan, 2016). We convert solar energy to either electrical or thermal energy for consumption.

In India, solar renewable energy segment is expanding at very fast rate because it is clean energy source, which is easy to deploy and operate.

11.5.6.1 Technology Trends

Operation and maintenance optimization is thrust area for this fast evolving solar panel technology. With increase in size and volume, it becomes imperative that emerging technologies are deployed to optimize solar power generation. Remote monitoring and automation of solar power plants can help in communication and monitoring of different components on a single dashboard (Gaikar et al. 2021). Remote operation centres can be deployed for operation management of solar power plants. Automated monitoring systems supported by AI and ML algorithms can enable predictive maintenance of panels, accessories and optimize solar power production. Drone-based surveillance of solar parks can enable better management of remote assets. Thermal infrared camera deployed on drones can detect defects in solar panels. Robots can be deployed for panel cleaning operations. They can also help in construction of solar parks. AI assisted robots are compatible with most structures, mounting areas and temperatures.

11.6 CONCLUSION AND FUTURE SCOPE

Energy demand and its consumption will increase with time. However, the conventional fossil fuels are limited and they pose a threat to our environment due to their high carbon footprint. The renewable and clean energy sources are the viable alternate option and switchover to clean and green energy sources will be a continuous phenomenon across globe. The share of renewables like solar, hydrogen, geothermal, wind, tidal and biofuels in energy basket will increase with time. The energy transformation will be a technology driven phenomena. Solar and biofuel have already captured some energy market share while hydrogen seems to gain pace with global impetus on green hydrogen production. The role of digital technologies in this transformation is very important as they fast track new solution development by providing insights and analytics from large data volumes and helping better management of transformation. The emerging technologies like AI, ML, big data, analytics, IoT, cloud computing have bigger role to play in this upcoming energy transformation. The role of emerging engineering technologies in the field of material, construction, design, environmental and allied areas is also critical to fast track transformation to clean and green energy.

REFERENCES

Abdalla, M. A., Hossain, S., Nisfindy, O. B., Azad, A. T., Dawood, M. and Azad, A. K. 2018. Hydrogen production, storage, transportation and key challenges with applications: A review. *Energy Conversion and Management* 165:602–627.

Alberto, B., Jamal, N. and Ayman, A. M. 2021. Hydrogen Production by Solar Thermochemical Water-Splitting Cycle via a Beam down Concentrator. *Front. Energy Res* 9:666191.

Anderson, A. and Rezaie, B. 2019. Geothermal technology: Trends and potential role in a sustainable future. *Applied Energy* 248:18–34.

Bhawangirkar, D. R., Nair, V. C., Prasad, S. K. and Sangwai, J. S. 2021. Natural Gas Hydrates in the Krishna-Godavari Basin Sediments under Marine Reservoir Conditions: Thermodynamics and Dissociation Kinetics using Thermal Stimulation. *Energy & Fuels* 35–10: 8685–8698.

Carella, R. 1985. Geothermal activity in Italy: Present status and future prospects. *Geothermics* 14: 247–254.

Chaulya, S. K. and Prasad, G. M. 2016. Slope Failure Mechanism and Monitoring Techniques. In *Sensing and Monitoring Technologies for Mines and Hazardous Areas*, ed. S.K. Chaulya, G. M. Prasad., Elsevier, 1–86, ISBN 9780128031940.

Chia, S. R., Ong, H. C., Chew, K. W., Show, P. L., Phang, S., Ling, T. C., Nagarajan, D., Lee, D. and Chang, J. 2018. Sustainable approaches for algae utilization in bioenergy production. *Renewable Energy* 129 B: 838–852.

Child, M., Kemfert, C., Bogdanov, D. and Breyer, C. 2019. Flexible electricity generation, grid exchange and storage for the transition to a 100% renewable energy system in Europe. *Renewable Energy* 139: 80–101.

Chowdhury, M. S., Rahman, K. S., Selvanathan, V., Nuthammachot, N., Suklueng, M., Mostafaeipour, A. and Habib, A. 2021. Current trends and prospects of tidal energy technology. *Environ. Dev. and Sustainability* 23: 8179–8194.

Craig, J., Absar, A., Bhat, G., Cadel, G., Hafiz, M., Hakhoo, N. and Thusu, B. 2013. Hot Springs and the geothermal energy potential of Jammu & Kashmir State, N.W. Himalaya, India. *Earth Science Reviews* 126: 156–177.

Dhyani, V. and Bhaskar, T. 2018. A comprehensive review on the pyrolysis of lignocellulosic biomass. *Renewable Energy* 129 B: 695–716.

Dimri, V. P. 2013 Geothermal energy resources in Uttarakhand, India. *The Journal of Indian Geophysical Union* 17 4: 403–408.

Dunne, J. A., Jackson, S. C. and Harte, J. 2013. Greenhouse Effect. In *Encyclopedia of Biodiversity Second*. Ed. S. A. Levin., Academic Press, 18–32, ISBN 9780123847201.

Fonseca, E. 2014. Emerging Technologies and the Future of Hydraulic Fracturing Design in Unconventional Gas and Tight Oil. IPTC, Doha, Qatar. 17439-MS.

Gaikar, B. V., Deshmukh, R. G., Rajasanthosh, T., Chowdhury, S., Sesharao, Y. and Abilmazhinov, Y. 2021. IoT based solar energy monitoring system. *Materials Today: Proceedings*, ISSN 2214-7853.

Gunnlaugsson, E. 2004. Geothermal District heating in Reykjavik, Iceland. *Proceedings in international geothermal days POLAND Zakopane*.

Haldar, S. K. 2018. Elements of Mining. In *Mineral Exploration* (Second Edition), ed. S. K. Haldar. Elsevier, 229–258. ISBN 9780128140222.

Johnstone, C. M., Pratt, D. Clarke, J. A. and Grant, A. D. 2012. A techno-economic analysis of tidal energy technology. *Renewable Energy* 49: 101–106.

Kannan, A. and Vakeesan, D. 2016. Solar energy for future world - A review. *Renewable and Sustainable Energy Reviews* 62: 1092–1105.

Kazmarczyk, M., Tomaszewska, B. and Operacz, A. 2020. Sustainable utilization of low enthalpy geothermal resources to electricity generation through a cascade system. *Energies* MDPI 1–18. doi: 10.3390/en13102495

Khare, S. K., 2019. A method and system for determining slip status of drill string. Patent US 10,066,473 B2.

Khare, S.K., 2021. Machine vision for drill string slip status detection. Petroleum Research. In press.

Koroteev, D. and Tekic, Z. 2021. Artificial intelligence in oil and gas upstream: Trends, challenges, and scenarios for the future. *Energy and AI* 3: 100041.

Ma, X., Gao, M., Gao, Z., Wang, J., Zhang, M., Ma, Y. and Wang, Q. 2018. Past, current, and future research on microalga-derived biodiesel: a critical review and bibliometric analysis. *Environ Sci. Pollut. Res* 25: 10596–10610.

Mohammadpoor, M. and Torabi, F. 2020. Big Data analytics in oil and gas industry: An emerging trend. *Petroleum* 6 – 4: 321–328.

Neil, C. W., Mehana, M., Hjelm, R. P., Hawley, M. E., Watkins, E. B., Mao, Y., Viswanathan, H., Kang, Q. and Xu, H. 2020. Reduced methane recovery at high pressure due to methane trapping in shale nanopores. *Commun. Earth Environ* 1: 49.

Nicolae, S., Dallemand, J. and Fahl, F. 2018. Biogas: Developments and perspectives in Europe. *Renewable Energy* 129 A: 457–472.

Rand, D.A.J. and Dell, R. M. 2008. *Hydrogen energy: challenges and prospects*. RSC Publishers. ISBN 978-0-85404-597-6.

Sain K. and Gupta H. 2012. Gas hydrates in India: Potential and development. *Gondwana Research* 22–2: 645–657.

Sharma, S. and Sinha, S. 2019. Indian wind energy & its development-policies-barriers: An overview. *Environmental and Sustainability Indicators* 1–2: 100003.

Stetco, A., Dinmohammadi, F., Zhao, X., Robu, V., Flynn, D., Barnes, M., Keane, J. and Nenadic, G. 2019. Machine learning methods for wind turbine condition monitoring: A review. *Renewable Energy* 133: 620–635.

Uihlein, A. and Magagna, D. 2016. Wave and tidal current energy – A review of the current state of research beyond technology. *Renewable and Sustainable Energy Reviews* 58: 1070–1081.

Vengosh, A., Warner, N., Jackson, R. and Darrah, T. 2013. The Effects of Shale Gas Exploration and Hydraulic Fracturing on the Quality of Water Resources in the United States. *Procedia Earth and Planetary Science* 7: 863–866.

Yadav, K. and Sircar, A. 2021. Geothermal energy provinces in India: A renewable heritage. *International Journal of Geoheritage and Parks*. 9-1: 93–107.

Index

A

B

C

D

E

F

G

H

I

J

K

L

M

N

O

P

Q

R

Y

Z

Printed in the United States
by Baker & Taylor Publisher Services